$29.95

My dear Emma

In 1895, Robert Emeric Tyler, a leading London architect, and his nineteen-year-old son Bobby travelled by rail and sea to Australia with a view to purchasing a goldmine. Their adventures, misadventures and incredible naivety as they journeyed in remote parts of the Western Australian goldfields were recorded faithfully in letters to 'My Dear Emma', Robert's wife in England.

Part travelogue, part social history, *My Dear Emma* is a remarkable personal eye-witness account of an Australia of one hundred years ago — of places, people and events from Coolgardie to Cue, Albany to Perth. The letters also provide fascinating insight into the world and attitudes of the late nineteenth century gentleman traveller, coming to Australia via Italy, Suez and Colombo.

Robert Emeric Tyler, 1907.

My dear Emma

A Full and Detailed Account of the Journey
of Robert Emeric Tyler and his Son,
to Western Australia,
and their Return to England.
August 1st 1895 to March 7th 1896

EDITED BY
ROBERT EMERIC TYLER IV

Fremantle Arts Centre Press
Australia's finest small publisher

First published 2003 by
FREMANTLE ARTS CENTRE PRESS
25 Quarry Street, Fremantle
(PO Box 158, North Fremantle 6159)
Western Australia.
www.facp.iinet.net.au

Copyright Introduction, Postscript and compilation
© Robert Emeric Tyler, 2003.

This book is copyright. Apart from any fair dealing for the purpose of private study, research, criticism or review, as permitted under the Copyright Act, no part may be reproduced by any process without written permission. Enquiries should be made to the publisher.

Consultant Editor Janet Blagg.
Production Coordinator Vanessa Rycroft.
Cover Designer Marion Duke.

Typeset by Fremantle Arts Centre Press
and printed by Griffin Press.

National Library of Australia
Cataloguing-in-publication data

Tyler, Robert Emeric.
My dear Emma.

ISBN 1 86368 352 6.

1. Tyler, Emma, — Correspondence. 2. Tyler, Robert Emeric, — Journeys. 3. Tyler, Robert Emeric, — Correspondence. 4. Architects — Great Britian — Correspondence. 5. Western Australia — Description and travel — 1851–1900. I. Tyler, Robert, 1941– . II. Title.

919.410432

The State of Western Australia has made an investment in this project through ArtsWA in association with the Lotteries Commission.

Introduction

On 1st August 1895, Robert Emeric Tyler F.R.I.B.A., a leading London architect and his nineteen-year-old son Bobby travelled by rail and sea to Western Australia, arriving at the Port of Albany on 31st August. Their intent was to visit the Coolgardie and Murchison Goldfields with a view to purchasing a mine or mines on behalf of an English syndicate.

Their adventures, misadventures and incredible naivety as they journeyed in remote parts of the state were recorded faithfully in letters to "My Dear Emma" and Lulu, Robert Snr's wife and daughter in England. Included with these letters were many press clippings, maps and photographs he thought would be of interest to family at home. To further explain things, sketches by Robert were also included with his letters.

Upon their return to England, these letters and sketches were copied into a leather-bound diary clasped with brass. At this time Robert also made margin notes in the diary to expand on things seen and done, and to include recollections not recorded in his original letters. The press clippings, photographs and maps were also glued into the diary adjacent to appropriate references in the text.

This book is a reproduction of the first half of this diary and includes a selection only of the sketches, press clippings, photographs and maps.

Robert Emeric Tyler was the Chairman of the South Londonderry Gold and Exploration Ltd. As the following letters record, the South

Londonderry Gold Mine proved worthless and its manager was prosecuted. Robert Tyler then purchased the First Find Mine near Bullabulling and this was operated by the First Find Consolidated Gold Mines, of which Tyler was Director.

The First Find was eighteen miles west of Coolgardie and operated until 1900.

Robert Tyler was also Director of The Murchison Gift Goldmining Company which purchased and operated the Polar Star mine in the Cue–Day Dawn area. A sample crushing returned 6oz of gold per ton of ore. The initial operation produced 4,000 tons of ore (worth $14 million in present-day terms).

Ultimately R. E. Tyler became chairman and director of Murchison Associated Gold Mines Limited. This company successfully operated several leases in the Cue and Day Dawn area into the early 1900s. Mines included the Rubicon, Polar Star and East Fingall, which were later managed by his son Bobby.

Little else is known about Robert Tyler Snr, or his wife Emma, except that he seems to have been an inveterate traveller, and that he died in 1908.

ROBERT EMERIC TYLER IV

My dear Emma

Robert Emeric Tyler, en route to Dunsville.

A Strange Country

"Australia, in many things, is a country in which nature has established conditions unknown elsewhere. It is a country full of contradictions and curiosities in animal, vegetable, and human life. Its native race, in point of intelligence and development of resources, is below the cave dwellers, and the people of the stone age of Europe. Its animals perpetuate types which disappeared from every other part of the globe millions of years ago. Its trees and plants are representative of species found elsewhere only in chalk and coal measures. Hardly any thing in that country has the character and quality of its relations in other lands. Although the trees and flowers are chiefly those of the temperate zone, the birds are for the most part of the tropics and flash the gorgeous colours of the parrot and the cockatoo, through the dull foliage of the sad-toned eucalyptus. The birds have no song, and such notes as they possess seem like weird echoes from a period when reptiles were assuming wings, and filling the tree tops with a strange jargon, before heard only in the swamps and ferns.

"The flowers have no scent, while the leaves of every tree are full of odour. (It is the only country that does not produce a wild rose.)

"The trees cast no shade, since every leaf is set at edge against the sun, and shed, not their leaves, but their bark, which, stripping off in long scales, exposes the naked wood beneath, and adds to the ghastly effect which the forest already holds, in the pallid hues of its foliage."

[Unsourced text copied as a frontispiece to his leather-bound diary by Robert Emeric Tyler.]

Emma, Robert, Bobby and Lulu Tyler, c. 1886.

Orient Line
"R.M.S. *Ormuz*"
Wednesday August 7th 1895

My dear Emma,

I hope you received the short notes which we posted up to Napoli, they were written just as "sort o' comforters" to let you know that we were well and in good spirits, reserving for a fuller letter a general description of our journey.

August 1.

You were of course told that we started well, the run to Folkestone was rapid and pleasant, and we quite enjoyed the voyage to Boulogne — the sandwiches came in most useful, but when we arrived at Boulogne the sight of the Buffet quite overcame us and we immediately sat down to "Poulet aux Cressons" and a bottle of claret, it did us good and sustained us until our arrival in Paris. It was a hot journey, carriage full, comprising three aged women, a Parson and a man with a bird cage, returning from Australia. When not asleep he was like "Count Fosco" with his white mice, always talking to his little birds in a confidential sort of way.*

[* Count Fosco is a villianous character from the novel *The Woman in White* by Wilkie Collins.]

Fortunately we had no trouble with our luggage as it was not examined either in Boulogne or Paris. At Paris we found we had about two hours and a half to wait, but the Lyons Railway from which we had to go from, being on the other side of Paris, beyond the site of the Old Bastille, we had to take a carriage so that we had a drive right across the City. Before however doing this we found a Café and had a real good "bif tek" aux pomines etc and some excellent wine, so we started well refreshed and in fine condition for Napoli. The Café we dined at was at the corner of the Rue Lafayette called the Café Montholon — we had time at the station before leaving to get a good wash and brush up — this took place in the Ladies Retiring room, and although several came in, our presence did not appear to disturb them in any way whatever — the custom of the country.

August 2.

The train started at 9.30, we slept until 5.30 and on waking found ourselves in sight of the mountains, the first Bobby had ever seen; his admiration was great. Mountains seem to impress you and make you think of the wonders of nature. We passed Dijon, the train running through a very pretty country by the Jura Alps, till we arrived at Modâne where our luggage was examined, and we had again to pay for it to Napoli, as at this point travellers are handed over to the Italian Railway Company, and they know how to charge. I have however been running on too rapidly. After leaving Dijon, we passed through Macon, Bourg, Culoz, Aix les Bains, St Jean, and then came Modâne.

Dijon.

Modâne.

At Aignebelle, a small village beyond Aix les Bains, we first saw snow on the mountains. By the way, before coming to Aix les Bains, we ran round Lake Bourget, very pretty, and passed the Termatt

Mountains. This lake is 15 miles long — the valley of Chambery was most beautiful, in fact all the scenery is beyond my description — we wished you and Lulu could have been with us to see it all, and enjoy it.

Modâne as you will see is at the entrance of the Mont Cenis Tunnel. We determined we would have a picnic in the Tunnel, as at Modâne we purchased rolls, beef and wine, which we thoroughly enjoyed, as we had only had a cup of Café since we left Paris, I think it was at Dijon. It took 27 minutes to get through the tunnel, but before getting to it you have to pass through two smaller tunnels, each taking from 5 to 10 minutes. We then entered Italy, and if anything the Alps on the Italian side are more beautiful than the French because they are less rugged and are cultivated high up. The view looking back at Mont Cenis was simply grand. The corn had been cut and stood in sheaves, on every little space up the mountain side stood the beautiful golden corn, the effect being lovely in the extreme. The Italians cultivate every yard that will take cultivation. On some of the ledges, on the very brow of the mountain, you could see two or three sheaves, seldom a patch of more than twenty or thirty together. The whole of the valley after leaving the Tunnel was under splendid cultivation, and the train instead of running in the valley, as in the French side, was on the side of the mountains, some distance up, so that the view was magnificent.

Mont Cenis.

Italy.

In this part of the country I noticed that they have stone walls dividing the fields, as in Ireland and

Cornwall; in France they have hedges as in England. As we neared Turin hedges appeared, and the country reminded us of England. I should have mentioned that the valley of St Jean before Modâne was most beautiful, and at St Michel, a station beyond, some rapids ran by the side of the line which were picturesque in the extreme, in fact all along the railway first on one side, then on the other, ran a mountain stream, at some places fairly wide. At La Praz we passed the highest mountain we had seen the peak of which was covered with snow. At Chambery we had hoped to see Mt Blanc but unfortunately it was not visible.

Turin.

At Turin we only got a glimpse of the Cathedral, which stands on high ground. After Turin we passed through Alessandria and Novi Ligure to Genoa, of course running through many other small places, but none of any importance.

Genoa.
Spezia.
Pisa.

The harbour of Genoa is very fine and we saw it well from the train. From Genoa to Spezia and on to Pisa the line runs by the side of the Mediterranean Sea; it was a clear moonlight night, and the views we got as we passed along were charming, something like from Dawlish to Teignmouth in Devonshire — unfortunately we fell asleep and did not wake up until we arrived in Pisa station, thereby missing the leaning Tower, which is a few miles out of the City; certainly we saw what looked like a huge chimney in the distance, but it was not the Tower. We saw by moonlight a little of Spezia, where the most beautiful white marble comes from, and where

August 3.

Charles Lever, the writer of *Charles O'Malley* and other Irish works, lived as Consul for many years. The line continued running by the sea, but through rather a barren district to Civita Vecchia, and then on to Rome.

Civita Vecchia.

Rome.

At Rome we had to change into another train for Napoli, but before the train left we had about an hour and forty minutes; we deposited our bags etc in the Cloak room, and rushed into a carriage — of course not speaking a word of the language we were rather nonplussed, but fortunately we ran across a Guide belonging to Cook's and he arranged all, instructing the Driver where to take us, and what time to return in order to catch the train. We saw Rome in one hour — at least we saw all we could in that time, of course only the outsides of the different places of note except one grand church which I have unfortunately forgotten the name of, but we have a photo of it at home. We saw the Colosseum, the Capitol, the Arches of Trajan and Constantine, the Forum and many other wonderful, interesting and ancient structures. In memory of Romulus and Remus, they still keep two wolves in a cage on the top of the Capitol; we drove to the top and from there the view of the excavations of the ancient Forum, the enormous columns, looking as if they had only been executed about one hundred years ago, instead of being buried in 16 to 18 feet of rubbish for more than 2000 years, was truly grand. We also saw the beautiful fountain of Neptune, and the King's Palace. St Peters and the Vatican we had a magnificent view of, but it was too far to drive in the time. In fact I guess, that in the time we saw as much of Rome as any other two men ever did.

We had only ten minutes to spare when we got back to the Station — Robert rushed for the luggage and I bolted into the Restaurant and purchased rolls, veal, a bottle of wine (Chianti, the especial wine of the district), a bottle of Roman water for 1 franc, very good, very cold and very dear, but greatly refreshing. I just managed to jump into the train as it was commencing to move off, but we did enjoy our lunch. You can imagine me running down the platform, a bottle under each arm, and holding the eatables in both hands. It was in this train that we met the young American who may call on you. Well, from Rome we passed on to Napoli, passing principally through mountains, or rather with mountains on one side, and very barren land on the other, that nearest the sea making it very uninteresting.

At one town, I think it was Caserta, we saw the King's Country palace; we also had fine views of Capua. What struck me very much was that all the towns we saw in the distance were built round high hills, with the castle or fortress on the top, no doubt for protection in times of war; evidently they had been so built from the earliest ages. Another thing struck us — the rivers were perfectly dry, wide rivers, with good stone bridges over, but no water. I suppose they were only for winter use; no doubt the water comes down from the mountains at the time when the snow melts and then they are raging torrents. I had no idea Italy

Capua.

was so mountainous. I should imagine the north of Italy is much more fruitful than the South — in the former you see any quantity of corn and other produce growing, but in the latter, the land seems poor, and they grow olives principally, and figs. It was a long run from Rome to Napoli in the very heat of the day, but still being interested in what we saw, everything being so new and strange, we enjoyed it.

Naples.

At Napoli we had to get our luggage out of the train, and as we were determined to see Pompeii, and as the train left in half an hour, we had to move about. I discovered another Cook's Guide, such a nice young man — spoke English well although an Italian. He rushed our luggage into the Depôt, lodged our bags in the Cloak room, got our tickets and landed us safely into the train for Pompeii: by the way, the ticket office had closed, but he managed to get them, and I paid him when in the train. It was sharp work, but we were not to be beaten by small obstacles, we were cool, like Phileas Fogg who went round the world in 80 days. To go to Pompeii you pass round one side of the Bay of Naples — to the left — quite close to the sea all the way.

Some one wrote — "See the Bay of Naples and die." That is all very well in theory, I have now seen it, and I don't want to die yet, but it is lovely, no one could but be charmed with it. All round the Bay are small villages with bathing arrangements, and charming Villas, where I assume the nobles and rich merchants of Naples pass their time, and then last, but not least, rises Vesuvius grandly in the distance. Pompeii stands

Pompeii Destroyed AD 79.

on the opposite side of Vesuvius; on arriving there we were besieged by drivers, who wanted to drive us to the ruins of Pompeii; fortunately we preferred to walk, otherwise murder might have ensued. We walked there in two minutes. The entrance has been carefully placed at the rear of the Hotel, so cannot be seen from the Pompeii Railway station. But before exploring the dead City, inasmuch as we had not washed for many hours, and your son looked more like a Stoker than a respectable young Englishman, we felt a good clean up would be advisable.

The ruins of Pompeii were discovered by some peasants in 1750, when digging in a Vineyard near the Sarno.

Having carried this out thoroughly and partaken of a good bottle of beer, we engaged a guide at one shilling and sixpence each, and commenced our inspection of the ruins: at the time of the eruption the City covered an area of four miles, and they have now excavated, and uncovered about two. It took us nearly two hours and a half, most fatiguing, but oh so interesting!

A Bronze Statue by the famous sculptor Polcrates, who was a a rival of Phidias, has been discovered. It represents an idol, and is valued at two million lires.

I could have spent days looking about it, and going over and examining the various houses and Temples, the Forum and other buildings. The streets are all quite narrow, just room for one cart or carriage to pass at a time with a pavement on each side. I should say the roads would be about 7 to 9 feet wide, and the paths about 3 feet, with a deep kerb of over a foot in height, and with stepping stones for people to cross the roads, sometimes two stones, and in the wider roads three; the side streets would be narrower than I have given. I imagine the roads ran with water hence the necessity for the stepping stones. At the side of the roads are large stone or marble cisterns about 4 feet by 3 feet, and 3 feet high, and the water flowed in — through the mouth of a head, or mask; you see the marble worn away where the drinkers placed their hands on the edge of the cistern to lean over to drink, and the side of the lips of the head are also worn away, showing that they must have been in use for an enormous time; from these cisterns the waters no doubt flowed into the roads. The idea being, I should say, to keep the City cool. The roads are laid with lava slabs, about one foot nine long, by one foot three wide, mostly laid diagonally, you can see the ruts in the roads, some of which are quite four inches deep.

Everything that was worth seeing in Pompeii we saw. The baths were most interesting to me, as I had not long back made sketches of the plans etc. The most interesting houses were those of Glaucus, Dromede and Lallust, and a more beautiful one, which had only lately been discovered, it must have belonged to a man of wealth and taste, as the paintings on the walls are very beautiful & so are the pavements in the various chambers. It is wonderful how they have been preserved, seeing that the City was destroyed A.D. 79, the colours have not even lost their brightness. You see the Baker's Shop, and the ovens out of which bread was taken and which you see preserved in the Museum. The wine shops with the large earthenware pots, or receptacles for the wine, standing in the very position they stood at the time, the money changers or Bankers, with the place on which the money was put down, and shops where various other Traders lived. You can see the leaden pipes laid in the Streets to supply the Cisterns, and in the Museum they have beautiful specimens of glass, brass and copper work, locks, jewellery and all sorts of implements. They were undoubtedly a highly cultured and artistic people, but not strong in their morals.

In the Museum you see the figures of a Gladiator; a male and female slave, having belts around their bodies; also a mother and daughter, all these were overtaken by the lava and killed, you absolutely see their teeth and bones protruding through the lava; there is also a dog. But the most beautiful things that were found have been removed to the Museum at Naples, some description of which I shall give later. On leaving I was presented with a flower which is still blooming in a glass in my cabin.

The last few lines I have written in my cabin, the former I wrote in the writing room. Bobby is now seated on my bunk, watching me write and giving me the benefit of his ideas. He is very well, and, you will be pleased to hear, has made friends with a very nice young fellow, who was at St Paul's School. They are always together so that he cannot be dull. The feeding in his class* is excellent — good breakfast — porridge, veal cutlets, tomatoes and potatoes, and two cups of coffee — dinner 1 p.m. Mutton and onion sauce, potatoes, cabbage and apple tart — High tea at 5.30 cold roast beef and jam and two cups of tea. Supper at 9. Bread and cheese — I call that good feeding. He drinks claret and water. I see him three or four times a day, and we walk together and chat about things generally, and of our friends at

[* Robert travelled first class at company expense, whilst Bobby travelled second class as the family was paying his fare. Robert also felt that Bobby was more likely to meet his peers in second class.]

home. But I have again digressed. On leaving Pompeii we again visited the Hotel at the entrance to the ruins, and partook of sandwiches and beer, then took a carriage to a station nearer to Naples, and caught the express back. Our guide met us, he got our luggage, and had it put on a carriage; four men were engaged in this operation, and I had to pay 3d for each article, and when we arrived at the Quay, more men had to be paid 3d each article for taking them off the carriage and putting them into a small boat in which we were rowed to the good ship Ormuz. I forgot to mention that when we arrived at Naples we were told that the ship had arrived in port, so we determined to go direct on board instead of making for a Hotel.

London to Albany via Gibraltar 10,288 miles.
Naples to Albany 7,970 miles.
London to Naples about 1,030 miles.

R.M.S. Ormuz. 6.031 tons, 8,500 H.P.

We arrived on the ship about 9 o'clock, somewhat fatigued for we had had a long and fatiguing journey, from London to Naples without stopping, besides the walking about the ruins of Pompeii. I came off well, but unfortunately Bobby came off badly. I had a good dinner and half a bottle of claret, but he only had bread and cheese. Of course I only heard of this the next morning, however, he was none the worse for it, and made up at breakfast. The following morning — Sunday — we left the ship before 10 o'clock to see Naples, and having secured a guide (he was blind with one eye and couldn't see far out of the other, however, he was not bad at 2/- for the day) we saw the Church of Jesus, the Cathedral, the Museum, and the King's Palace. By that time, about 2 p.m., he was as limp as a wet rag — it was hot — we had given

August 4.

Officers on the Ormuz: Capt. Tuke, 1st Officer Mr Fairly, 2nd Officer Mr Kershaw, 3rd Officer Mr Taylor, 4th Officer Mr Bryant, Purser Mr Fox, Chief Engineer Mr MacInnis, Doctor.

Registered at Stationers Hall.

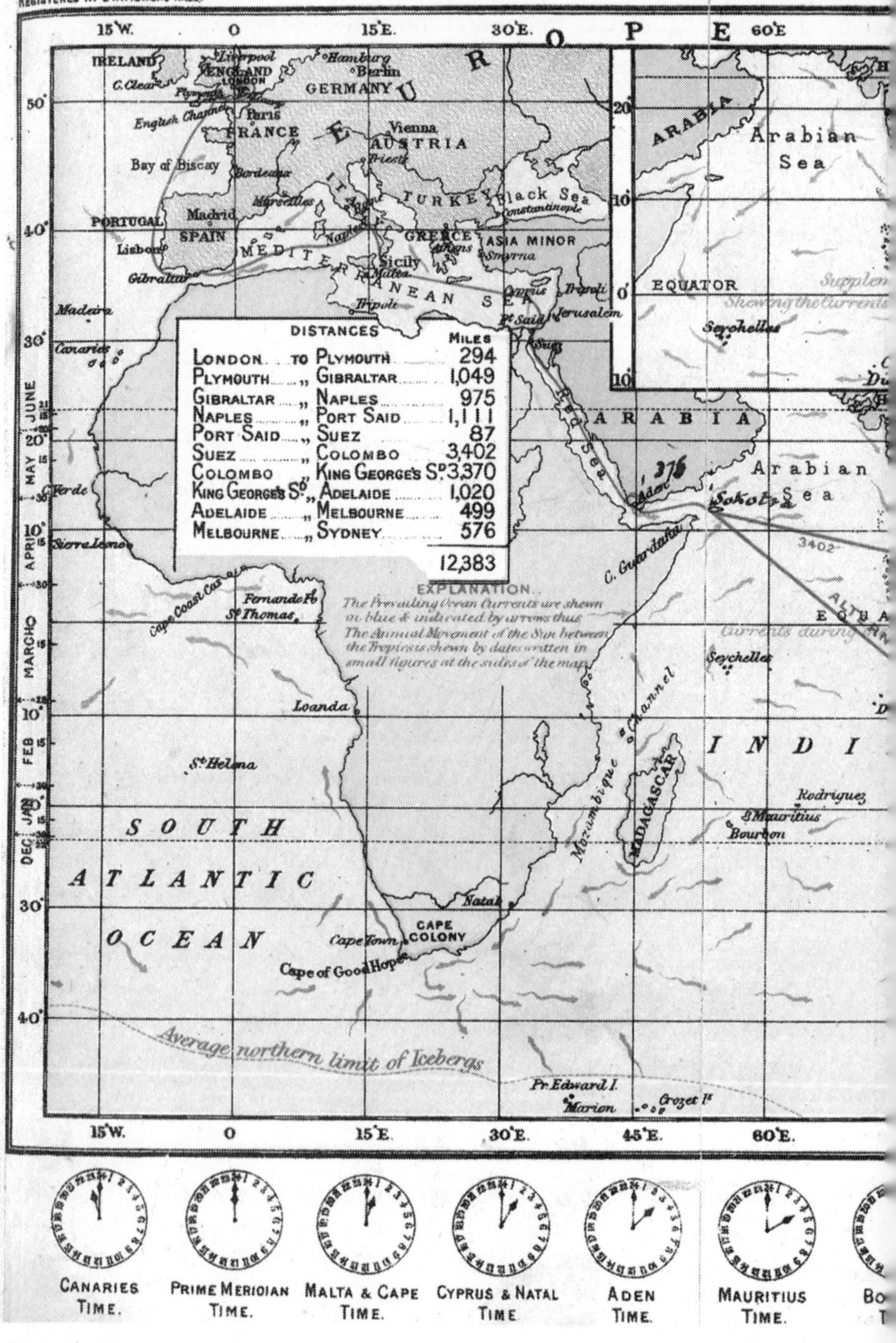

ORIENT LINE PASSENGERS' T

DISTANCES			Miles
London	to	Plymouth	294
Plymouth	"	Gibraltar	1,049
Gibraltar	"	Naples	975
Naples	"	Port Said	1,111
Port Said	"	Suez	87
Suez	"	Colombo	3,402
Colombo	"	King George's Sd.	3,370
King George's Sd.	"	Adelaide	1,020
Adelaide	"	Melbourne	499
Melbourne	"	Sydney	576
			12,383

EXPLANATION.

The Prevailing Ocean Currents are shewn in blue & indicated by arrows thus. The Annual Movement of the Sun between the Tropics is shewn by dates written in small figures at the sides of the map.

Canaries Time. | Prime Meridian Time. | Malta & Cape Time. | Cyprus & Natal Time. | Aden Time. | Mauritius Time. | Bo[...] T[...]

…IND. AND CURRENT CHART.

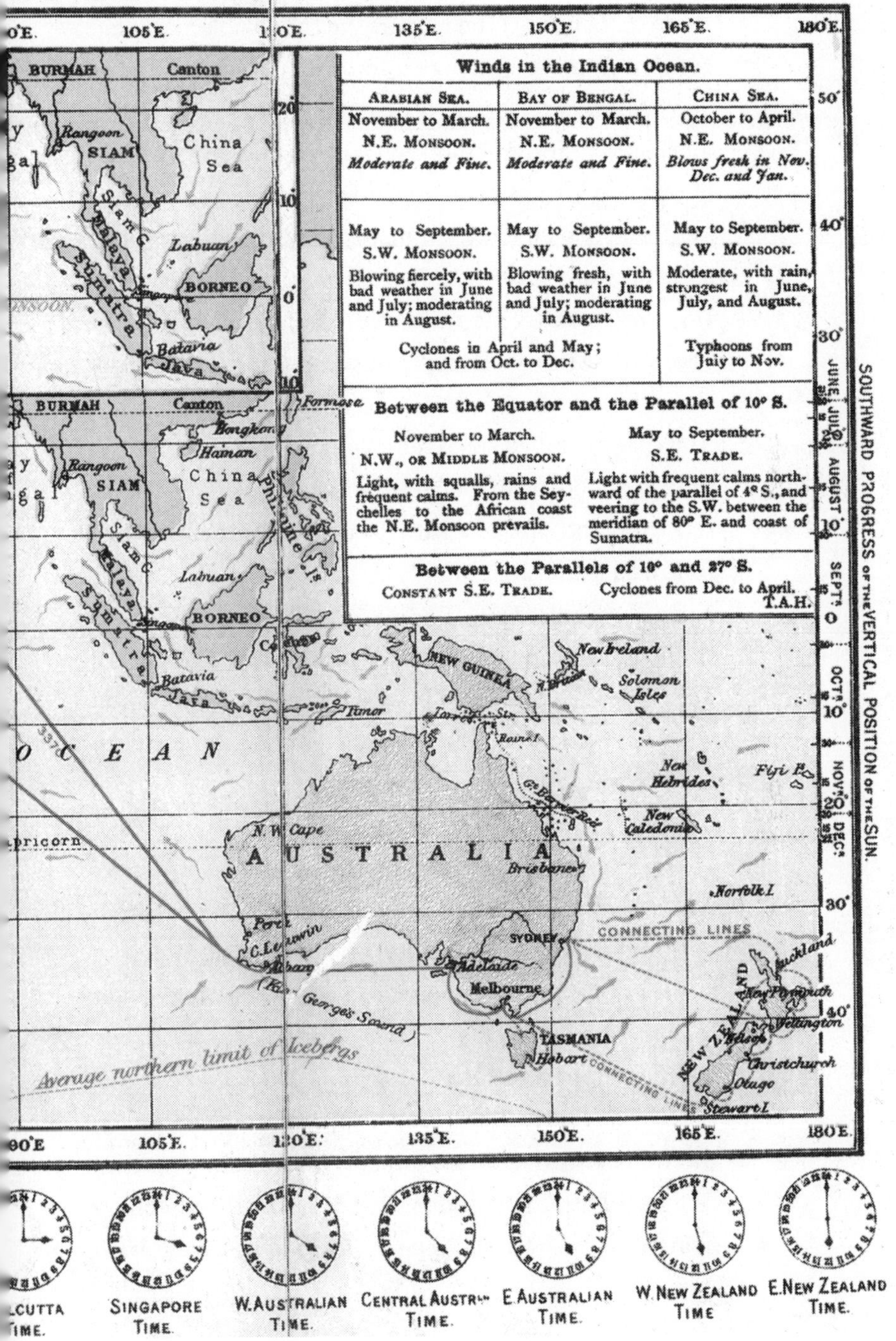

Passengers on the Ormuz: Father Murphy, Father Cahill and two other priests, Col. Kilgour and wife, Mr Canning M.E., Mr Frank Ravers, Mr Lane, Dr Howith, Miss Cruddas, Mr Figg R.N., Mr Green R.N., Mrs Riley, Mr Robert Brough, Mrs Brough, Myles Bircket Foster, Mr Brown.

him wine and beer, which he was unused to probably — so he begged to be let off. We then took a carriage for two hours and drove round the high points of the city, where we obtained beautiful views of the Bay, and surrounding country. Afterwards we walked about the streets — it was a fete day, every one out — until six o'clock, then made up our minds to dine, anything but a pure Italian dinner would suit us — this we obtained at the Gran Café de l'Universe in the Via Roma — It certainly was good. Maccaroni (small) aux Tomátes — Red mullet with large Maccaroni, Fillet de Boeuf cooked in oil, beautiful potatoes and beans, cheese and fruit, one bottle of red wine and one of white, large, 4 francs each, then followed Café and brandy — all very good. About eight we went to see the Sea Fete, three of men war, ships all illuminated — a fine effect and all sorts of boats decorated with Chinese lanterns, winding up with fireworks. At eleven we caught the tender back to the ship, having spent a very interesting and pleasant day.

The streets of Naples are not by any means wide, and the side streets remind one of the rows at Great Yarmouth, but a little wider, about 10 to 12 feet; the houses are high, and there are no pavements even in streets of 20 to 30 feet.

Church of Jesus.

The Church of Jesus is all built of lava, about 250 years old, the ceiling beautifully painted by Rafael, the walls are covered with inlaid marbles in beautiful designs. The Cathedral is also beautiful, the ceiling painted by Michael Angelo, the walls inlaid with marble, the general effect being simply lovely.

The Museum is the great sight of Naples, and is said to contain the finest bronzes in the world — the sculpture is also very grand. The Farnese Bull consists of the Bull, five figures, and a dog, all life size, carved out of one block of white marble. The bronze horse brought from Pompeii is magnificent. The Gladiator also from Pompeii. The drunken Bacchus, The Runners and many other beautiful works of the ancients. Unfortunately, it being Sunday, no guide books were sold. The collection of glass, and other things found at Pompeii and Herculanum, would alone take a week to properly examine; besides there are many paintings, by the greatest Italian Artists, magnificient works of art. When we grow rich we will all go and spend some time there.

The Museum.

The Farnese Bull.

The King's Palace was very grand — not so fine as Buckingham Palace but still very fine, the rooms well proportioned and lofty and some very fine paintings. I have written all this today, in fact bar meals and a little rest after dinner, I have kept at it, as the mail is supposed to go out tomorrow from Port Said. It is hot, but my cabin is on the windy side of the ship — the Port side — so that's all right, with port hole open, I am comparatively cool. I have had heat bumps on my hands but not more than I used to have at the seaside. I use Vinolia cream and I think it is doing me good. I had a talk with the Doctor about it, he thought nothing of it, only suggesting that meat once a day was enough, so I am following his advice. I do not eat much, of course enough, but in hot

The King's Palace.

climates you can really do with less food. In health I am perfect, so is Bobby; he is, I am sure, enjoying himself thoroughly, and looks ever so much better than when he left England. Were it not for the cockroaches I should be absolutely happy. I spend about half an hour before getting into my bunk hunting them up, and I am generally successful in slaughtering a few.

Now how are you and Lulu getting on! We trust you are having a good time at Margate, and that the change is doing you both much good; we shall expect to see you both looking splendid when we return. You will give our united love to all the families, and friends, who take an interest in our movements. We often think of all our friends from whom we are parted, but it will be a great pleasure to return and relate all we have seen, as it is impossible to describe everything in a letter, although I have done my best. I am now writing on my trunk, and the time is 12.40 so I think it is time I said good night, and God Bless you. Our next letter will be posted either from Suez or Colombo.

I forgot to mention that one of my fellow passengers is Mr Brough — a nephew of Lionel Brough and cousin to Sydney who lives opposite to Nellie. I have not yet interviewed him, or his wife, who is very charming. All told, I fancy we are only 25 in the first class. None up to much, only one young lady, with her companion, she is but a poor thing — about twenty, can't sleep, and uses the clubs to develop her chest. There are four Roman Catholic

Fathers, to see them playing at cricket in their long black coats is more than amusing.

One afternoon at three o'clock I was called up by the Captain (I was then writing) to see "Venus". I suppose she does not often appear in the afternoon (too shy I imagine); however I look, and look again through the Captain's glass, but was unable to discover her.

Venus.

Your affectionate Husband
Robert Emeric Tyler

R.M.S. Ormuz.

August 9, 1895

My dear Emma,

Just a line to say that we have arrived at Suez all well. The run through the Canal has not been anything like so hot as I had anticipated. In fact on deck last night — I remained up until about 3 — was beautifully cool.

Your affectionate Husband
Robert Emeric Tyler

The Gulf of Suez — Red Sea
August 9, 1895

My dear Emma,

I posted a short letter this morning at Suez just to let you know that we had passed through the Canal. Before commencing an account of our journey from Naples to Colombo, the next posting point, I will give a few further particulars of Pompeii and Naples which may interest any readers (I think that's a neat way of putting it, quite in the novel writing style.)

The Farnese Bull in the Museum was brought from Rhodes to Rome before the time of Pliny — it was sculptured from a single block of marble, the Artists being Appollonius and Tauriscus — 2nd century BC — It is now in many pieces and its restoration was principally carried out by Michael Angelo. The subject is Antiope intervening in favour of her rival Dirce [Circe?], whom her sons are about to bind to the horns of a wild bull. The Farnese Heracles, by Glycon of Athens, is another magnificent work, also restored by Michael Angelo. In the Museum you also see the well known bust of Homer, the statue of Antinous, the supposed statue of Aristides, the Venus of Capua, the Gladiator, and the lovely Psyche formed in the amphitheatre at Capur. In the picture gallery we saw some fine Correggios, one being the brilliant "Marriage of St Catherine" — Titian's portrait of Philip the Second (magnificent) and Raphael's

The Farnese Bull.

In the house of the Grand Duke of Tuscany in Pompeii was found a painting of this. It was originally cut out of a single block, at the base, 10ft x 10ft.

The Farnese Heracles.

Bust of Homer

"Madonna del Divino Amore".

In the Pompeii section you see chairs, tables, bread, eggs, combs, toothpicks, surgical instruments, painted vases, iron stocks and loaded dice (besides not being strictly moral they were up to a little bit of swindling). Most of the bronzes were found at Herculanum not Pompeii, the former was buried in lava, the latter in ashes.

August 4.

In my former letter I believe I said that we went on board about 11.30 on Sunday night. I waited up until the mails arrived, about 12.30, but after seeing more than one hundred bags of mails hoisted onto the ship I had had enough of it and retired to my little bunk. This was a small mail, 600 bags, each bag quite sufficient for one man to carry — 10 bags are brought up from the tender at a time, they are put onto an octagonal shaped cloth with ropes at each angle, the ropes are then placed over a large hook, then hoisted by crane onto the deck, and carried to the strong room, which is fire-proof.

August 5.
Capri.
Ships run 118 m.

In the morning I was told that we left Naples at 5.30: we passed close to Capri, the highest point of which is 1,980 feet above the sea. Unfortunately we had not the time whilst in Naples to visit the famous Blue Grotto, or the remains of the Villas of the Emperor Tiberius — he had, it is said, 12 palaces on this Island — evidently he was fond of a change of scene. Of course we had a fine view of Vesuvius, which is 4,100 feet above sea level.

After Capri the ship passed into the Gulf of Salerno, then came the elevated headland of

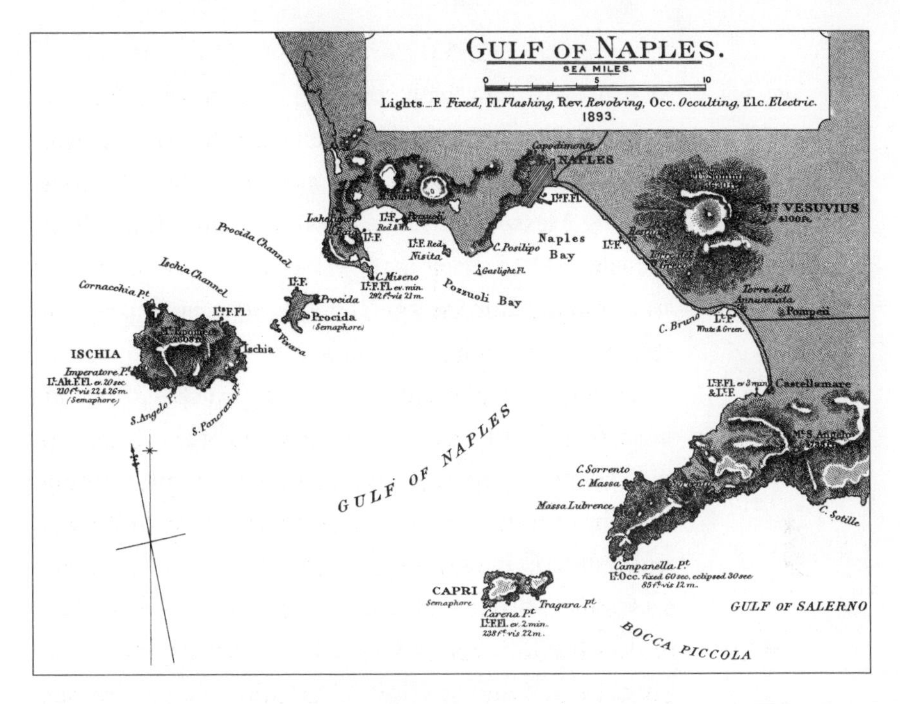

Palinuro, named after Palinurus the pilot of Aeneas. And then Stromboli which is still an active volcano. To the Westward the Lipari Islands were passed. One small islet volcano has of late years been very active. Two hundred miles from Naples commence the Straits of Messina; we passed between the rock of Scylla and the whirlpool Charybdis, and close to Cape di Faro, the Pelorum of the Ancients. The first you see of Sicily is the light house built by the English during their occupation at the commencement of the century. On the shores of Sicily you see Messina, founded by the Greeks 3,000 years ago, and still a flourishing port — on the Italian shore you see Reggio, the capital of Calabria, a town apparently of some importance — again you see

Straits of Messina.

Reggio.

many rivers perfectly dry, but no doubt, at certain seasons, they are well supplied with water from the mountains, the mountain range runs through the centre of Italy, down to Cape Spartivento. After passing Messina we had a good view of "Etna" which viewed from the sea is very grand. There is no other mountain in Europe which gives the same impression of height — it is 10,880 feet above the level of the sea — Syracuse, once the largest of Greek Cities, we passed but could not see. In the days of the Greeks it was 14 miles round, having a population of hundreds of thousands, whose wealth and luxury was unbounded; it is now a provincial town of some 19,300 inhabitants. It was founded in 734 B.C.

Syracuse
August 6.
Ships run 350 m.

On the Italian coast the rocks are very fine, being rough, steep and precipitous. Behind Reggio are the wooded heights of Aspromonte; Garibaldi was defeated and taken prisoner by the Italian troops on the plains below, on the 29th August 1862.

August 7.
Ships run 350 m.

After passing the straits of Messina the ship steamed straight for Egypt; on the left we passed Candia or Crete, the mountain peaks being in view for some hours; we also saw the lighthouse at Crete, which is the highest in the world.

Port Said.

We arrived at Port Said on the 8th at 9.30 a.m.: at 10 o'clock we landed and spent three hours looking about and inspecting the town. Port Said is built on the sand not two feet above the level of the Mediterranean and is very uninteresting, no end of beggars and touts of all descriptions. To a stranger it is naturally somewhat interesting, the Arabs in their picturesque dress, the streets and houses, women with their faces covered up, all but the eyes; Negroes, thieves, cripples, beggars, sellers of everything you can imagine, more especially photos, which are very cheap, but we did not purchase any as we can do so on our return journey. We walked round the town, and inspected the Arabs quarter. There were plenty of donkeys and camels, but we declined a ride into the desert.

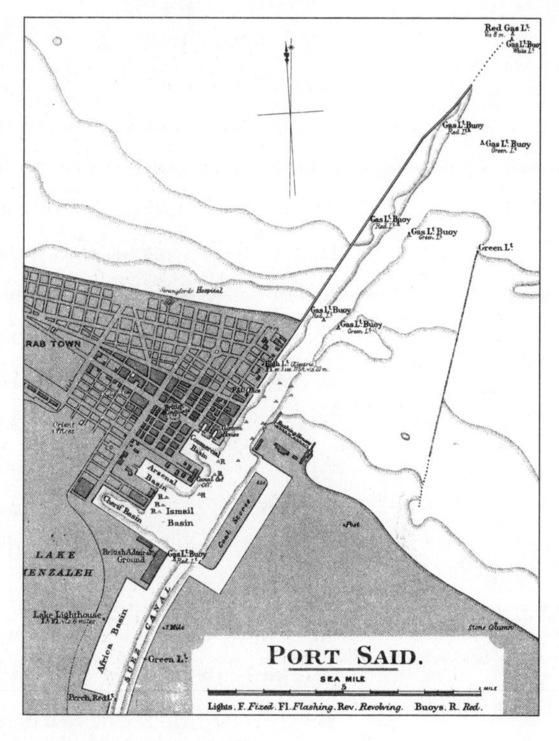

Here we coaled — it was quite a sight to see — three barges fastened together were hauled on each side of the ship; on the barges, beside the coals, were about 200 Arabs, as black as the coals which they put into baskets, and carry on board. I suppose we took in over 1000 tons — it was most amusing to watch the Arab boys diving for pence, which they caught before the coppers reached the bottom of the sea.

Port Said is the entrance to the Canal built on the low sandy shore of the Delta; in the distance we saw Damietta and the entrances to the Nile. The Canal Company's offices is a fine building, and the Hotel is not bad, but how any one can live there on the borders of the desert — I can't understand, it is only fit for flies, and camels.

Suez Canal.

We left Port Said at 1 p.m. and steamed up the Canal. I should mention that the distance from Naples to Port Said is 1,111 miles, and from Port Said to Suez — that is the length of the Canal — 100 miles. The view from the steamer whilst passing through the Canal is by no means picturesque but it is interesting, on the left or east, towards Jerusalem, is a sandy desert as far as the eye can reach — on the right is Lake Menzaleh with its swamps and marshes. In the distance could be seen hundreds of flamingos and pelicans. We were told they were flamingos and pelicans, but they were a long way off, however, a few we could see. Here we watched for a mirage — certainly I saw what looked like a lake, where no such thing was, but I did not see any ships upside down. Some ruins of the time of Rameses are to be seen in the desert not far from the Canal. At the 44th Kilometre, Al Kantarah — which means arch or bridge — is reached; at this point, a hill was cut through. "It is by this natural bridge," says Dr Wallis Bridge, "that every invading army must have entered Egypt," and its appellation "the bridge of Nations" is most appropriate. It is still the most frequented route across the Canal into Syria, and a kind of wooden draw-bridge crosses the waterway.

Lake Tinusah.

Between Al Kantarah and Lake Timsah (crocodile) 17 miles of narrow canal with high banks have to be traversed — possibly crocodiles at one time existed in this lake, hence the name. At the commencement of this part of the canal, are some difficult curves, and the pace is slower. The general rate of speed is from four to five miles an hour. The cutting through this part was the heaviest piece of work, the ridge of Al Guior rose 70 to 100 feet above the level of the desert. This is the one strip of terra firma between Lake Timsah and the marshes of Lake Menzaleh. All along this strip of land runs the ancient desert route to Syria.

I should mention that a fresh water Canal runs by the side of the Canal, and at Lake Timsah turns abruptly away to the West, through the land of Goshen, to Ismailia, where the summer palace of the Khedive is situated.

About 9.30 p.m. yesterday, we were in Lake Timsah making for the cutting which goes through Joussoum to the North entrance of the Bitter Lakes, a sharp curve at this point has to be made, and the French Pilot, who came on board at Port Said, let the ship run out of her course, which is marked by a number of buoys — the engines had to reverse which caused her to tremble to such an extent that every one rushed to ascertain the cause; to get back into proper position, the anchor had to be dropped. This caused the sand and weed at the bottom of the Lake to be disturbed and the stench was something awful. The result being a delay of more than an hour, but eventually she was got back into position, and steamed ahead.

Possibly Pharaoh and all his hosts, may have been destroyed at the particular spot, where the unpleasant odour arose — who knows! From Lake Timsah to the north entrance to the Bitter Lakes is about nine miles, each side being desert.

The Bitter Lakes.

The Lake widens out considerably, but the channel is still narrow — about 30 yards — buoys being placed 330 yards apart to mark the waterway. These Lakes, there are two, The Great Bitter Lake, and the Little Bitter Lake, are the remains of a dried up, or partially dried up, arm of the Red Sea, on which once flourished the ancient port of Arsinoe — To the west is seen the verdure line of the fresh water canal, that now runs from the Nile to Suez. Behind this rises the Geneffeh range of mountains or hills. Getting out of my evening dress, and putting on my wester and slippers, I remained on deck watching the scenery we were passing through; about twelve feeling somewhat fatigued I sat down on a deck chair to rest, fell asleep and did not wake up until two o'clock, the moon was then shining brightly, and it was cool and beautiful. I found that signals had been made, that a ship was coming through the narrow part of the canal that connected with the Gulf of Suez, so our steamer had to drop anchor. I remained up until 3 then turned in until 7.

August 9.

From the Bitter Lakes to Suez the banks of the Canal are hard, and at Chalouf the cutting is through sandstone, after that to Madama it passes through sand hills, marsh and clay; just before coming to Suez it again passes through sand hills. As the steamer enters the Bay of Suez, after leaving the Canal, you see the desert that has always been associated with the wandering of the Israelites, and Mount Attallea rises boldly to a height of 2,700 feet. As a matter of fact the Canal does not go to Suez, the port is three miles from it, the official name being Port Tewfik, but locally it is called Terreplain.

We arrived at Suez about 8.30 and should have left within an hour — time only being allowed for the Company's Agent to come on board, and letters to be taken off. We started, but it soon became evident that something was amiss; the Captain wore a worried look, so did the officer and crew, the ship stopped, after a time it leaked out that a hawser, that is a rope about 4" diameter — had become wound around the propeller. An Arab diver was fetched from the shore, and he and the second officer dived down, taking knives with them to cut the hawser, but it was no good, so the Captain sent for a professional diver, who in due

Suez 8.30 a.m.

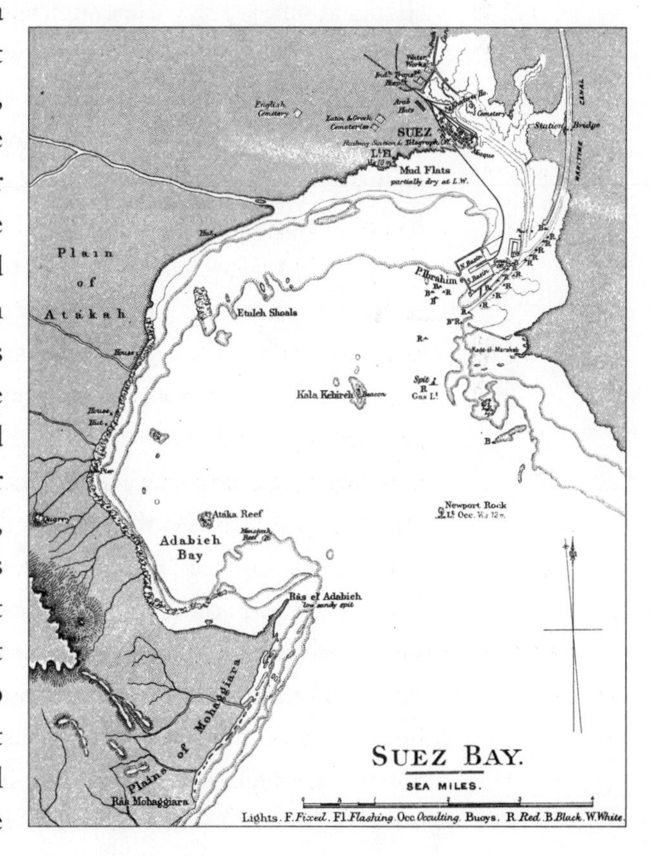

course arrived with his apparatus. The diver was hard at work cutting out the hawser — a specimen of which Bobby has, until midnight; he kept down for an hour at a time coming up for a rest of 20 minutes. All this time, which we might have passed on shore, for it was within a mile — we mooned about discussing the position, and watching as much of the operation as we could, which was not much — then night came on, an electric lantern was rigged up and let down, so that the diver could see what he was doing. We were told that there were any quantity of sharks about; we saw none, neither did they attempt to attack the diver. The accident was annoying, a day being lost, but they say it will be made up before we arrive at Colombo. Whilst at anchor we saw two caravans — one with 200 camels, the other about half that number.

August 10.
Ships run 154 m.

At 12.55 a.m. we again started. I remained up until about 1.30 then had a cup of coffee and to bed. This morning there is a most pleasant breeze, and although the sea is rougher than we had yet had it, I enjoyed my breakfast — haddock, curried fowl, two poached eggs, a green fig and two glasses of iced coffee. I have seen Bobby, he is well and jolly with his young friend — we are not quite out of the Gulf of Suez, said to be about 165 miles long; on each side are ranges of high hills, sort of small mountains, running down to the edge of the water, but no verdure is to be seen, not a tree or shrub — it is now 12.45 and we are not yet out of the Gulf. On leaving the Bay of Suez we passed, to the East, an oasis of tall palm trees

Gulf of Suez.

known as Moses' Well — then an isolated mountain known as Jebel Sudur — sailors call it the Saddle Mountain. It was here that Professor Palmer and his companions were murdered in August 1882. The Gulf is from 10 to 25 miles in width, on each side mountains, table lands and bold ridges from three to six thousand feet high, they are most picturesque in outline, and the rich and varied colouring is marvellous. To the east is the Peninsula of Sinai, at the southern end of which is Jebel Katherina 8,630 feet high — (12.15 p.m.).

We have now entered the Red Sea, it is hot, but not unbearable, ices have just been handed round on deck — very good. The heat bumps I have been suffering from are better today, by tomorrow I think they will have departed — I can well do without them — The writing room I find as cool as any part of the ship, there is no glare.

The Red Sea.

Sunday 12.10 a.m. it is hotter — This is my first night in the Red Sea — on deck there is a cool breeze, but down here — in my cabin — it is sultry, but still not hotter than I expected. I am remarkably well, but then I eat carefully, and also drink in moderation, enough, but not too much. We are now out of sight of land so I have not much to write about. I was very interested in watching the flying fish, once I saw at least 30 at one time; they, I understand, fly out of the water when the Dolphins are after them, but the latter I did not see, neither have I yet seen a shark, I hope to. The flying fish seem to be about the size of a herring, 6 to 7 inches

August 11.
Ships run 336 m.

long. We are getting on. We have now seen Arabia and Africa — that is to say little bits of these. I had a conversation with one of the Roman Catholic Priests on the subject of Moses in the wilderness. My view is that the wilderness was a huge forest, not the barren desert it is at the present time; he quite agreed with me. Why! Moses and the Israelites would not have lasted a month, they would have been fizzled up.

Whilst writing my hands are covered with beads of perspiration. The Purser — a pleasant man who sits at the head of my table — has just looked in, I suppose to see if I have melted away. I enjoy the sea bathe and shower every morning, and the sponge down with fresh water after. I dress for dinner, as it is cooler to be in your dress clothes, and besides it is something to do. The ship is making good way now, and will make up for the time lost at Suez — And now, good night and God bless you — 3.10 walked the quarter deck until 4.30, then returned to bed, a fine night. 7.30 have just had my bath, and seen Bobby who is quite well. The Church service took place on deck at 10.45 — The Captain conducted the service, and the Chief Officer (Mr Fairly) read the lessons — it was most impressive. Mr Foster accompanied the Hymns on the piano — most of the first class passengers attended and some of the second, amongst them being Bobby and his friend. The Roman Catholic priests, I assume, had a service to themselves, anyhow they were not present at our service. After church the men on the ship were formed in line on the deck and inspected by the Captain, each man answering to his name. It has been awfully hot today, the water running off me, but still on deck there is a pleasant breeze. The sea has been like a lake all day, and the sunset was magnificent, such as you see in Turner's pictures. We are now going rather slower on account of the 12 Apostles, they are small islands or rocks, and ships only run near there by daylight. I slept from 12 to 5 then walked the deck for half an hour, then slept again until seven o'clock.

August 12.
Ships run 354 m.

August the 12th. It is hotter than yesterday, but we are told that directly the ship is out of the Red Sea it will be cooler. I am longing for that time to arrive — still I stand it well. Bobby was all right this morning, but last evening he felt the motion of the ship slightly; they get more of it at their end than we do.

From yesterday noon, to today noon, the ship did 354 miles, the fastest run this trip. The temperature of the sea was 84° and it is 104° in the shade. Fortunately my Cabin is on the Port side — that is the left — and though hot during the early part of the day, is cool at night. I have perspired awfully during the day, and have changed three times. I find we have not come to the Apostles yet, but expect to tomorrow morning. I walked over to see Bobby this evening. I found him quite jolly with two other boys; had he been with me, he would not have enjoyed himself at all.

August 13.
Ships run 355 m.

August 13th. Had a fairly good night — heat and cockroaches interfere somewhat with one's comfort;

just had my hair cut — temperature in Barber's Shop 94°, expected to be up to blood heat during the morning. This is the hottest day, the sooner it is over the better. It is now 8.30, and I am looking forward to iced coffee for breakfast. I take three tumblers.

Straits of Babul Mandib.

12.10. We are now just entering the Straits of Babul Mandib (the gate of tears), so we are out of the Red Sea. There are two ways out, the narrow way is called "Hell's Gate", between the two is the island of Perim, on which a lighthouse and telegraph station stands.

In a few minutes it will be known in England that the Ormuz has safely passed this point. The before-mentioned Apostles must have slipped by, or rather we must have slipped by them during the early morning. All I can say is, the Red Sea has behaved most kindly towards us, perfectly calm and lovely weather, bar the great heat. I have been on the look out for a Nautilus but have not yet seen one. A party by the name of Brown said he saw two. A small bird called a "Hoo-poo" from the noise it makes, has been following the ship for two days. I only saw it this morning, the 14th. It is about the size of a partridge — very pretty.

August 14. Ships run 362 m.

Gulf of Aden.

The change from the Red Sea to the Gulf of Aden did not quite agree with me — when dinner came I could only eat a little, but before bed time I felt much better, and this morning I was all right and enjoyed my breakfast, but when the bad weather comes I fear, more especially if we run into a Monsoon, that I shall be bad. I have changed into

August 15. Ships run 370 m.

the opposite cabin, as the Purser thought I should get more air. My former cabin is marked A on plan, my present one B, it is slightly larger. Last night I slept in the upper bunk immediately on a line with the porthole. My steward had arranged to provide a ladder, but he forgot it, however I managed to scramble up, and slept until 4 a.m. when I went up on deck for half an hour. I then slept until 7.45 so I had a good night's rest.

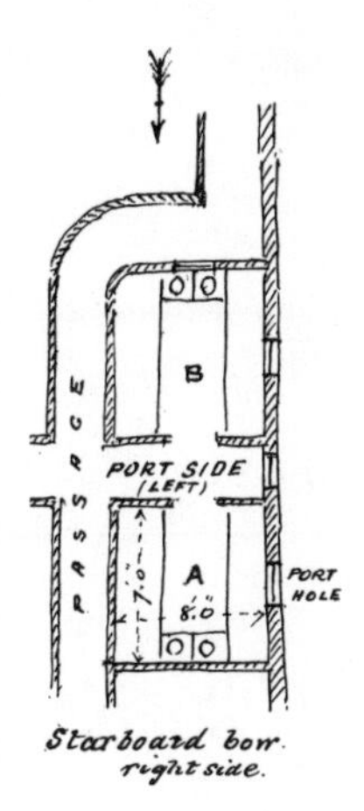

Starboard bow. right side.

After dinner last evening I was on deck, but not feeling well, and wanting my wester, I sent one of the sailors for Bobby, who came across and then fetched it for me. I have not seen him this morning.

August 16. Ships run 362 m.

This is Friday the 16th. Yesterday I could not take up a pen — at 7 p.m. on the 14th we struck the Monsoon — some stupid person called it the mongoose. I say we struck it, I consider it struck us, after that it was all up, the sea raged and roared, and swamped the deck, so that we had a bad time. I held out until 7 p.m. the following evening. "I saw a gentle priest, drop gently in a chair. I heard him heave a heavy sigh, I knew he felt it there." I felt it there at that hour, I was sick, but not much, the time is now 4 p.m. and I have not been ill again. I managed to eat a little on deck. I had chicken and tongue, washed down with stout, this was my luncheon, and I enjoyed it. Bobby has been very ill, but he is now quite well, and on deck enjoying himself. I had a chat with him before luncheon today. From what I can gather we are passing out of the Monsoon, so the sea is not as troubled. Very nearly all the passengers have

The fiddles appeared on the dining table, which is always a sure sign that dirty weather is expected.

been, or are ill, which is some consolation. We are now looking forward to Colombo, where we hope to have about nine hours on land, it will be a delightful change. I was sitting on deck yesterday, feeling far from well, the second officer was next to me, one of the young men passengers — there are only two — came up bothering, wanted to know what he could do, oh! I said, go and be sick — it was perhaps unkind, but when you feel sea sick, you don't want to be bothered.

August 17. Ships run 360 m.

Saturday the 18th. This has been an uneventful day, nothing to see but flying fish. In the afternoon the second class came across and played the first at cricket. Bobby was one of them, it was an exciting game, six on each side — one side made one bye, the other one run. In the evening they had dancing in the second class, as well as music, and singing. Bobby was dancing and thoroughly enjoying himself, in fact he is having a real good time. I am not sure if I said anything about Aden, it is a military station, not at all popular, and I hear it is considered a beastly hot hole. After Aden we passed Socotra, an island about 82 miles long. It was between these two places that the sea was at its worst — "we suffered there" — directly we came to the Island it commenced to calm down, and we all began to feel happier. I must say that the Indian Ocean has not behaved badly, on the contrary.

Socotra.

August 18. Ships run 359 m.

This morning, Sunday, the 19th I have eaten a real good breakfast — red mullet, devilled chicken, Cambridge sausages and ham, washed down with two tumblers of iced coffee, finishing with an orange and iced water, not bad I fancy you say. Well you know, you must support nature. This night I slept in the Smoking Room which is on deck. I was up at six, and by six thirty had had my bath and interviewed Bobby, who was quite up to the mark. At 10.45 church — a Clergyman took the service, he is a third class passenger, and I am told he is a good man, and gives an interesting sermon. I ascertained that the Holy Romans had their service at six, they are very jolly priests, four of them — in the morning they read the prayer book and Bible for two or three hours, they must know every word by heart, but still they have to keep at it. Perhaps they discover something new each morning — who knows! There is a very nice elderly Lady I chat with, she is a Holy Roman, but not strong at it, she can't believe in miracles. She is a sensible woman, certainly she is a plucky one, not less than sixty, quite white hair, has a husband and grandchildren, was determined to see the world, no one would come with her, started by herself, been round the world, through America and Europe, wished to go to Jerusalem, and would have gone, only Cook declined to take her — what do you think of that? Is now on her road back to the land of her birth to join her family, and intends to see Jerusalem next year. Now I don't suppose I shall have much more to say before we reach Colombo, which will be tomorrow at 3 or 4, so I shall bid you adieu with fond love and kisses. My secretary does not improve, he knows I can't sack him on board ship, this morning he absolutely declined to make up my washing list.

Ever your affectionate Husband,
Robert Emeric Tyler

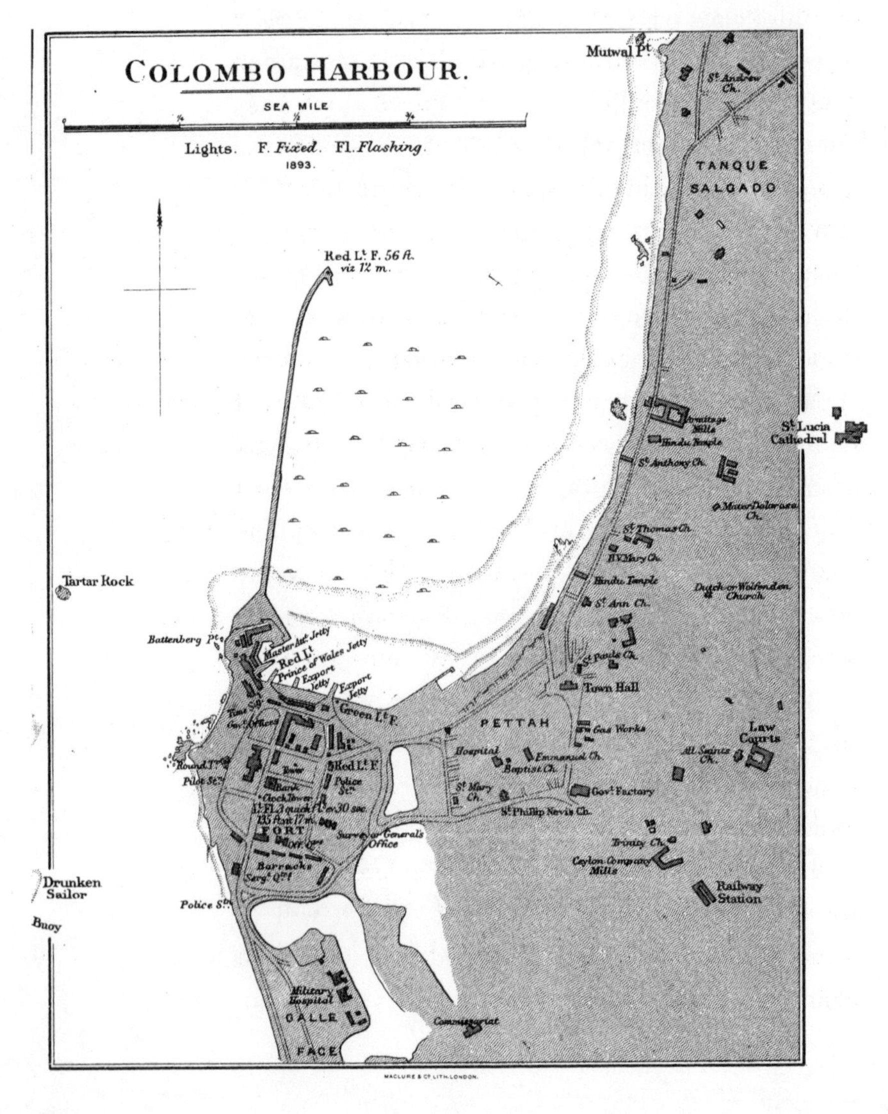

The *Ormuz* — Indian Ocean
August 20, 1895

My dear Emma,

I completed my last letter on Sunday morning, but dated it the 19th instead of the 18th, the stupid fellow in charge of the writing room had put us on a day, hence the mistake. We had service on the open deck, but this time a Clergyman was produced from the 3rd class, it was very impressive but he gave no sermon, however in the evening he preached on the 2nd deck. I went over to hear him but he did not please me. Bobby came to the service with his friend Orchard, such a nice clean looking boy, and really a good one.

August 19.
Ships run 346 m.

Monday the 19th. We are all in a state of excitement, nothing spoken about but the prospect of passing a few hours on land. Give me land. I don't want to ever be on the sea again; where the enjoyment comes in, I can't see. Not even the officers like it, and I'm sure the sailors don't look as if they did. Of course it is magnificent and grand, but soon becomes uninteresting and monotonous, to see the vast expanse of water for days without a sail on it, not even a bird to be seen, only occasional flying fish, well, I really don't think even a "Poet" could stand it for long. "Life on the Ocean Wave" — Bosh "Life on the flowering land" is my idea of true happiness, or where true happiness — if it is to be found on this globe — should be discovered.

Well, to proceed. Unfortunately we did not arrive off Colombo until 3 p.m. and it was 4 before we landed. We were delighted to get ashore, to feel you could walk about without being uncertain where your next foot step might land you. Mr Burket Foster (Professor Music), Bobby and I kept together — in fact did Colombo in each other's society. First we went to the tea kiosk and partook of tea and cakes — good tea, shall buy a case when returning — then went across the road to the Grand Oriental Hotel and secured seats for the Table d' Hote. Having done this we walked to the Post Office and posted your letter, the general letter Bobby had arranged with his steward to post, then we took a carriage and drove for two hours round the town. We were driven through the Cinnamon Gardens, round the fresh water lake and through the native quarters, seeing the markets and the shops of the natives as we passed along. It was most interesting. Of course we had never seen an Eastern Town or City before and we were delighted. The effect on landing from the Tender was at once surprising — red roads, perfectly white buildings, natives — those that were dressed — for some were very scantily attired — were dressed costumed in all sorts of bright and startling colours, the Jinrickshaws (these as you know are the two wheeled carriages drawn by natives) and small carriages with striped cotton awning over them, all formed a delightful and fascinating picture. You felt as if you could sit down, sip iced drinks, and simply watch the scene.

Colombo.

Colombo is a large City, 120,000 inhabitants and the Cyngalese, judging from what we saw, are a contented and happy race, they look well fed and satisfied with their lot. Besides which they are by no means bad looking. Some of the men are handsome and well built, whilst the women — we saw many walking about — were not at all unpleasant to look upon. One thing, they are upright, that is, they have straight backs, which arises from the custom they have, from early youth, of carrying articles on their heads. In colour they are light and dark bronze, with black hair, held back by a comb. When starting for our drive, a little native boy jumped up on the back of the carriage by the side of Bobby, he must have been about sixteen, such a nice looking little chap, and spoke English fairly well, he said he had picked it up entirely through talking to travellers. He acted as our guide, told us all about the different places we drove through, and picked flowers for us, beautiful flowers. Bobby and Mr Foster decorated their hats with the flowers and leaves.

Main street.

Native bungalow.

We saw any quantity of coconuts on the trees, and pine apples growing wild, all sorts of flowers and fruits with names we could make neither head nor tail of. Some parts of the road were beautifully shaded by magnificent banyan trees (*Ficus indica*) and other parts by the graceful Areca (*Areca catechu)* and the cocoa nut (*Cocos nucifera*). The Cinnamon Gardens were highly interesting. All cinnamon trees, with undergrowth of flowering shrubs, the drive through was about 4 miles which brought us round by the fresh water lake, on which small steamers run, so it must be of considerable extent. It is somewhat irregular that a fresh water lake should be within a few hundred yards of the ocean. It is a grand drive on the red road, by the side of the sea — it is called the "Galle Face" — magnificent waves rolling in from the vast ocean, how I hate that ocean — and to feel the bracing air it quite revived us, we felt happy and gay, in fact we were so, we laughed and talked and enjoyed the lovely scenes we passed through, little thinking of the morrow, that we had to pass on that boundless ocean, but more of that anon, sufficient for the day etc. We were on land — Oh blessed land — our hearts were full of joy, it wanted only you and Lulu to make the picture perfect.

Bobby wanted to buy some little presents for some young ladies — who with their Mama had been very kind in giving him afternoon tea, so our little black, or rather bronze guide, took us to a shop where he purchased bangles and sleeve links, very pretty and cheap; they were I believe highly appreciated, but causing much jealousy amongst some of the other fair damsels. I only purchased one article, a silver buckle for a waist band, I think Lulu will like it. I did not intend purchasing anything until the return journey, but it really seemed worth the money. I bought it from a native dealer, who had a kind of stall in the hall of the Hotel; I should say, he had his wares spread out on the marble floor. He offered it to one of the young fellows (Frank Travers) for two pounds, I said I thought the value of it would be about ten shillings, at which the dealer was quite indignant, but he gradually dropped to twenty five shilling, through one pound, and as I was leaving the Hotel ran after me into the roadway and consented to accept the price I had valued it at. They are awful thieves. Travers bought a ring for 30/- which he had been asked £7 for, so you see it wants care in making purchases in Colombo.

The dinner at the hotel was simply excellent, and with it we had (Bobby and I) a bottle of well iced Irroy champagne — very good — we did enjoy it — no rolling no tumbling or tossing of the sea (the beastly sea), it was peace.

After dinner we had coffee and cigars in the Verandah or Hall, soft and dulcet music being played somewhere. We knew not where, but it was simply soothing to our nerves, the Intermezzo was beautifully played, Mr Foster was delighted with it. But all joys come to an end, alas too soon. Time was up and we had to make for the ship, and such a ship, the finest in the world, when being coaled, is a thing to be avoided — it is simply covered with coal dust, every thing you touch is coal dust, you absorb it into your lungs, into the pores of your skin, in fact all over you, your clothes are filthy and everyone looks the picture of disgust and misery. Coupled with this all the port holes are closed, for two reasons, one to keep the coal dust out, the other to prevent the native gentlemen, who surround the ship in small boats, from hooking things out of the cabins; the port holes being closed the cabins are unbearable, so it was no use thinking of turning in. It was four o'clock before I turned in, and the ship left about 6 a.m.

On approaching Colombo you see a great number of small boats, called catamarans, they are singular, you wonder how it is they are not knocked over by the heavy waves, they are about 25 to 30 feet long — standing about 3 feet 6 inches out of the water, but in width not more than 8 to 10 inches, not room for a man to sit down, they are kept up by two poles, about 12 feet long, which project from the side of the boat, and these poles are connected with a baulk of timber — or tree — nearly as long as the boat; by this arrangement the boat is balanced, and she can ride safely in the roughest sea.

Breakwater.

The breakwater is about the finest in the world; the Prince of Wales laid the first stone in October 1875, it was designed by Sir John Coode. The view of Colombo from the sea is very fine, it is hilly and the coconut trees grow down to the shore, whilst bungalows, picturesquely situated, are seen dotted amongst the palms, they seem to be constructed of timber with red tiled roofs. All the time I have been writing, the rehearsal for a concert, which is to take place on the 29th Thursday, is going on, which is somewhat confusing.

I am writing this on Tuesday — until yesterday I had not taken up a pen since the 20th.

One of the sights in Colombo harbour are the Divers, they are all boys, it is wonderful how they dive and bring up the coins that are thrown before they get to the bottom, it was about 45 feet where we were. They have only a rag round their loins, they kneel on a raft, which is composed of three pieces of timber fastened together, so that directly a coin is thrown, they drop their paddle, with which they propel the rafter, and dive into the water. It is done in an instant, they watch the coin and by the time it touches the water, they are on its track. Five divers are on each raft, which is about 2 feet wide by 20 feet long. Each boy has a paddle and the one at the end guides, or rather steers, with his. To see these boys, practically naked, of different shades of bronze from dark to quite light bronze tint, is most interesting; when they emerge from the water the effect is fine, they really look like bronze figures, and they all have nice looking intelligent faces.

I have endeavoured to give you as interesting an account of Colombo as I possible could, but my pen has failed to do it justice.

At Colombo a Mr Lane left us, which I regretted

At Colombo they took on board some fruit called "Mangosteens", most delicious. The shell is brown and hard, but the inside is white and soft, they are about the size of an orange.

very much, he had been at the Cambridge, had turned planter, and purchased an estate called Blair Athol – after the horse that won the Derby in 1864. I was present at that race, a man won a lot of money on the race, and with it purchased this Estate, and Mr Lane purchased it from him. He commenced by growing coffee which eventually turning out a failure, he took to tea, some of which I hope to bring back with me, as a sample of the best Ceylon.

Blair Athol also won the St Ledger. Was owned by Mr L'Anson.

August 20. Ships run 102 m.

On the 20th we first had rain, and on the 21st we had another brush with the Monsoon (bad scratch to him). It rained hard, and the sea became very rough, but it was a grand sight — one I never desire to see more. At 7.15 p.m. we crossed the line, I am pleased to say nothing extraordinary occurred. They did have some fun on the second deck, which Bobby will tell you all about. For the first time I saw the "Southern Cross", very fine, these stars can only be seen on this side of the line.

August 21. Ships run 327 m.

August 22. Ships run 318 m.

Thursday the 22nd. Not up to much, felt not very well, played at deck billiards to pass away the time, rained heavily during the afternoon. Tropical rains are unlike our gentle showers. In the evening went to the concert given by the second class — it was not bad.

August 23. Ships run 279 m.

Friday the 23rd. Nothing occurred, went in for a sweep, the run of the ship — put 2/- in for Bobby, the same for myself, we both lost.

August 24. Ships run 276 m.

Saturday the 24th. Very rough in the "Trades". Much prefer "professions", not so breezy. I eat well, but felt considerable uncertainty as to what the result might be. Someone has just finished "Go pretty rose". It reminds me of Gower Street and Lulu.

August 25. Ships run 272 m.

Sunday service in the Saloon — by the third class Revd gentleman — he reads well, but can't preach. I understand he is aware of that fact, he has been a missionary in Mozambique. I attended a lecture he delivered last evening on Africa, and he told rather an amusing tale. The Sultan of Zanzibar was in the habit of sending presents to the mission. One day a cavalcade arrived — a very imposing one — bringing a present. After undoing various boxes, they came to a clock which the Sultan desired should be placed in the Chapel, this was duly carried out. Service was then held, but in the midst of it, the clock, it being a musical one, played "Sally in our Alley". The clock remained in the Chapel, it would not have been policy to annoy the Sultan.

August 26. Ships run 306 m.

Monday. A most beautiful cool pleasant day. Cricket was played on deck. In the evening the lecture before mentioned. I regret to say the Parson is in bad odour with his class, the third; they consider he should have lectured to them, not to the second, consequently he has had to apply to the Purser (Mr Fox) for protection. He is being hooted, jeered at, and otherwise insulted; they have not yet taken to rotten eggs, probably because they are not obtainable. The purser could do nothing, so he has to put up with his unpleasant position.

August 27. Ships run 335 m.

This afternoon the 2nd Class sports took place on our deck — your son immediately sprained his ankle in the sack race, so had to stand out.

August 28. Ships run 332 m.

August the 28th. This morning Bobby is nearly right again. I am delighted to say that we have now only two more days to be on board this ship — we expect to be at Albany on Saturday morning. I shall take the train straight for Perth — about 300 miles — I am told, and takes twenty four hours to do it in, at the rate of about 13 miles an hour, it seems absurd. I think this will amuse Percy. I quite enjoyed myself before dressing for dinner last evening. I commenced packing. It is awfully annoying that I should have left the map of West Australia behind, I suppose it was found in the letter press. I have wanted it so much, but I may be able to get one in Albany or Perth. Bobby has told you all about the sports that were held this afternoon, he did not distinguish himself.

August 29. Ships run 324 m.

August the 29th. This is about the rollingest day we have had. I don't like it a bit, but still I stand it, and eat well. I am closing this letter today, as tomorrow will be devoted to packing. We expect D. V.* to be in Albany on Friday night or early Saturday morning.

[* D. V.— Deo Volente, *God Willing*.]

With love to all kind friends and relations of every degree —

I remain, Your affectionate Husband,
Robert Emeric Tyler

Grand Hotel, Perth, W.A.
Sept. 5, 1895

Dear Mater,

As we have arrived at our head quarters and there is a good deal of business to be done I have taken on the correspondence. I think the Pater left off at the night before we sighted Albany.

August 30. Ships run 339 m.

On the Friday morn we packed — put all the clothes we wanted into the Pater's box and loaded mine with the rest and left it, with the chairs in the Custom House — which was a closed shed at Albany. In the evening there was a smoking-concert on our deck in which we (those who were getting off at Albany) had our healths drank and were wished good luck etc etc.

Albany.

We arrived in the harbour of Albany at about 11.30 p.m. and got on shore at midnight. The parting on board was very severe as I had been rather a favourite, in fact a few tears were shed. I received an invitation to visit people at Melbourne and Sydney.

King George Sound, Albany, Western Australia.

OYSTER HARBOUR
L. Seppings
Strawberry Hill
ALBANY
Middleton B.
Cheyne Hd.
Gull Rk.
Ledge Pt.
Mt. Adelaide 410
King Pt.
Red Lt.
Frederick Pt.
Lockyer B.
Bramble Pt.
Possession Pt.
Earker Bay
KING GEORGE
SOUND
Michaelmas Reefs
PRINCESS ROYAL
HARBOUR
Geak Pt.
Rushy Pt.
Stuarts Hd.
Pagoda Pt.
South Spit
Mistaken I.
Seal I.
Frenchman Bay
Flat Rk.
Limestone Head
Sharp Pt.
Green I.
Cave Pt.
Black Head
Stony Hill
Peak Head
Vancouvers Ledge

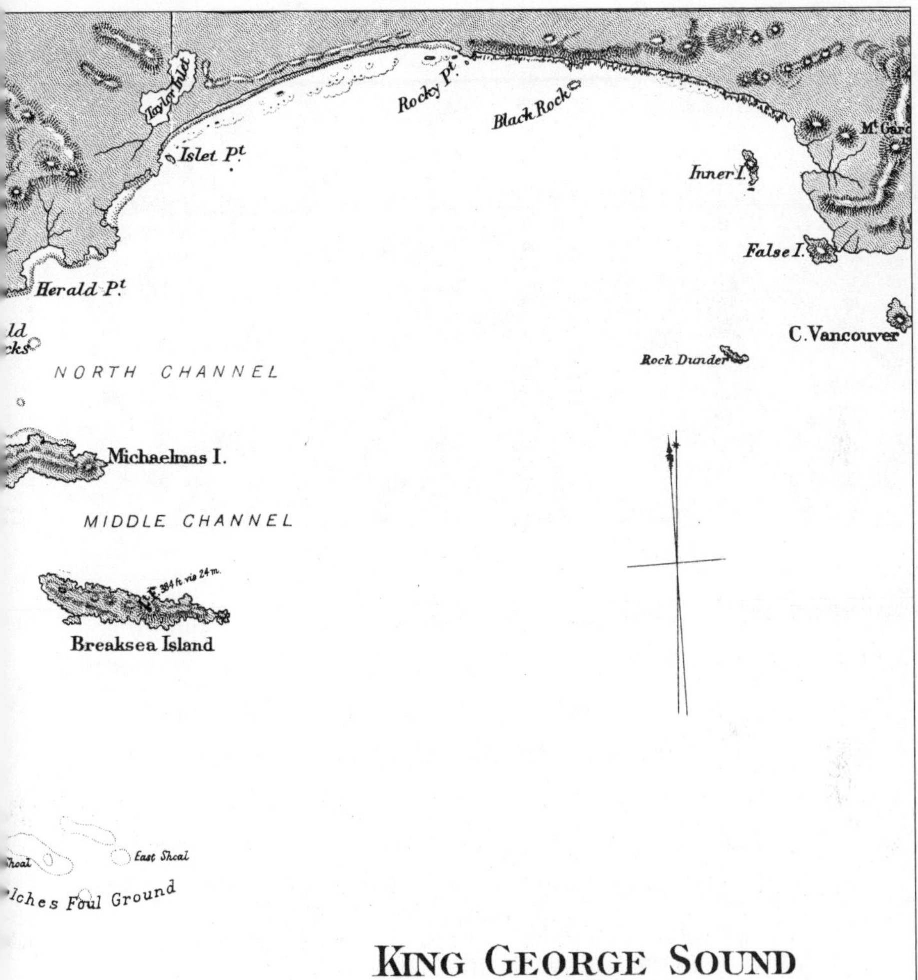

King George Sound and Princess Royal Harbour.

SEA MILES.

Lights _ F. *Fixed*. Fl. *Flashing*. Int. *Intermittent*. Rev. *Revolving*. Bn. *Beacon*.

1893.

Albany, from the jetty.

Talk about it being warm in Australia. I put on an extra vest and my overcoat, and was then simply freezing. The Pater met Mr Fitzgerald Moore, so I did not see him until the tender was leaving the Ship's side — when on it we found that it was the Government tender, meant for the mails only, no luggage — so nothing — so we had to wait on the jetty, in the bitter cold, for an hour until the next tender came. We then collared our luggage and made for the "Royal George Hotel", a fine place. The two Orchards and a fellow named Bird went with us. Bird got on at Colombo, had been tea planting there — very strong — holds the Amateur Championship weight lifting of England — 560 lbs — we call him Hercules. The following morning after a good

Royal George Hotel.

breakfast we strolled about — the air very cold — just like Margate in the Winter — very small place, about three streets — walked to the top of Mount Clarence — 633 ft high — climbed a flag staff and wrote my name. The Princess Royal Harbour, on which Albany is situated, is about 5 miles long by 2½ wide, land locked, the entrance being circuitous and narrow, on the Princess Royal Harbour opposite shore is a long low range of hills. The view from Mount Clarence over the harbour, the country round and the Ocean, or more properly speaking, I should say King George's Sound, for you see that first, was very grand and we were quite delighted. After a good dinner, we went to bed — the first bed we had slept in for a month, and slept well, not having been used to climbing, and not having walked much for some time, we were tired out.

August 31. Weighed on Albany Pier 12.6. Last weighed at Burton. 12.13.

Had "skipjack" for breakfast, a very nice fish.

Sunday September the 1st went to Church — a very good sermon — and at 5 o'clock started for Perth — the Orchards and Bird with us — pretty little engines but don't go badly — about 20 miles an hour, it is a narrow gauge line about 3'6" — we had violin solos (both the Orchards are very musical & play well) and played whilst stopped at Katanning, halfway to Beverley — and had food — quarter of an hour — managed to stow away a lot in the time — very uneventful journey — saw little log huts and tents — raining hard, arrived at Perth at 9 in the morning, a 16 hours run, 351 miles. Had been raining for a month.

September 1.

Albany to Perth 351 miles.

Albany to Beverley 241 miles.

The Grand Hotel that we are staying at, is one of

September 2.

the best in the City, and very comfortable, the feeding being particularly good.

September 3 Perth.

Perth is a fairly big city, but in 1891 the population did not exceed 9,617 — males 4,978, females about 360 under that figure — now it numbers about 20,000. It is situated on the Swan River, which is very wide at this particular part, but not sufficiently deep for large sized trading ships to come up from Fremantle, which is the port of Perth. There are no amusements here, but everybody seems to have lots of money. The Orchards have secured rooms, so we don't see much of them, Hercules however hangs with us still. The business is very complicated. We expect to leave for Coolgardie in two or three days. I am waiting to pick up a nugget, but have not done so yet. We are awfully fit and eat like horses. The Pater

Perth to Fremantle 12 miles.

Fremantle.

dined with Mr Horgan our Solicitor last evening, unfortunately I was out so missed him and the dinner besides. I talk with anyone I can get hold of and am full of information. They all try the pumping game, but it does not work with me. I arrange about the rooms, and the household work, as it were, the Pater looking after the business. Taking it altogether we had a grand voyage, and only missing one meal is a jolly good record, I think.

The Weld Club, Perth.

Made honorary member of the "Weld Club" proposed by W. Salmon, Manager of The Bank of New South Wales. The Weld Club is the best club in Perth called after Sir F. Weld one of its most respected Governors.

Gained 6lbs during the voyage and scale 10 stone, and the Pater has lost 7lbs. This shows the condition we are in, we are longing to hear from you. There are an enormous number of pretty girls here, in fact every girl you meet is pretty, they lick the English girls into fits.

Fremantle, the port, is not a bad place 3/4 of an hour's run from here by rail. The principal things here are whisky, bad language, cards & Billiards. We shall have to buy horses up at the mine, £3.10.0 each. Give my love to all relations and friends.

Ever your loving Son
Robert

Royal Hotel, Coolgardie
Sept. 13, 1895

Dear Mater,

Coolgardie stands 1682 feet above sea level, and is 350 miles from the Moore River.

Coolgardie was discovered by Arthur Bayley and William Ford, Sept 1892.

We got your letters today, and were very pleased to hear you are getting on alright. We are having a fine time, except that the business is a bit complicated, but we are sifting it out slowly but surely. Coolgardie is a jolly fine place in the day time, but at night there is absolutely nothing to do. We are very comfortable here and both well and strong. We are sending you a photo of ourselves by the next mail so look out for it.

Now I will commence my diary. I wrote to you last Wednesday the 4th. After I had finished writing we lunched, then the Orchards and Bird started with us for a walk to Mount Eliza which is very high ground, to the right of Perth, overlooking the Swan River. I believe it was christened after the first Governor's wife. It was a lovely day, and the view from the top of the Mount looking over the

The Swan River and Mill Point, South Perth, from Mount Eliza.

River with the Canning Bridge in the distance, which spans that part of the river which leads to Fremantle, was very fine.

More properly speaking I should say "Swan Waters", it is really a very large lake, through which the river passes on its way up to Guildford; and the high road to Fremantle runs by the side of the lake, passing at the base of Mount Eliza. This a very pretty drive, in fact the only good one out of Perth, as far as we saw. On this drive there are banana trees growing, the leaves being of huge size. The top of the Mount is covered with trees, scrub and sand. It is called the "Public Park of Perth", but up to the time we visited it, the only claim it had to the name of "Park", was from the fact that the Mayor of Perth had opened it and planted a small tree — a very small one — in commemoration of the same. Returning we called on Mr Horgan and I was introduced to his wife and daughter. I was then shown round, and later on introduced by Miss Horgan to her pony. In a rash moment I volunteered to ride it, the bridle was put on, but not the saddle; I mounted, the animal cantered into the garden, then I suppose I did something which annoyed and irritated it, and it commenced to "buck", I stayed there about the 1/8 of a second, was then dislodged and found myself on my back on the ground, much to the amusement of the spectators.

Mount Eliza.

It was a very nice house built in the bungalow style, with a verandah all round it, consisting of about 8 rooms, the kitchen built out at the rear, and separated from the house by the verandah. It has a flat roof, from which a fine view of Perth can be obtained. I should have mentioned that the house is situated on a part of Mount Eliza and stands in about 2 acres of ground. One of the advantages of a flat roof is, that in the hot weather, and it is hot there in the summer, sometimes up to 106° in the shade, you can take your bed onto the roof and sleep in the open. Our friend Mr Horgan has a fine telescope which he fixed on the roof, through which he inspects and studies the stars, and heavenly bodies.

September 5.

The following day, Thursday, it was much warmer, we purchased various things we required to take up with us to the fields, and did our packing, only taking with us what we considered would be necessary for a short stay. The rest of our things were packed in my trunk, which we determined to leave at the hotel.

September 6.

Perth to Coolgardie 366 m.

Perth to Northam 78 m.

Perth to Southern Cross 248 m.

Railway to Southern Cross opened 1st July 1893.

Rean's Soak to Coolgardie 98 m.

Friday, started for Coolgardie by the 3.30 train — Bird accompanying us. — about 8.30 we stopped at Northam for dinner, next at Parker's Road at about 5 a.m. — saw the sun rising, a lovely effect. Here we partook of some light refreshment — at 6.30 we arrived at Southern Cross, had breakfast, and started again at 9.30 for Rean's Soak — 20 miles. The line not being completed the train only went at the rate of 3½ miles an hour, very slow and weary travelling. The line had only been laid as far as Rean's Soak, so here we had to take the coach, constructed to hold 14, but on this occasion there were 17 beside the Driver. The Pater had a seat behind the driver, but my seat was amongst the luggage on the top, a most perilous position. The coach was drawn by 5 horses, 3 in front and was supported on leather springs. This is absolutely necessary on account of the state of the tracks to be driven over, of course you will understand they are not roads in any sense of the term. The bush and trees, had just been cut down in the first instance, to allow sufficient space for the teams, coaches etc to pass along, and had gradually become widened out, as the ruts became too deep. At "Hunts Dam" we had dinner, it was rough, but we did ample justice to it, then on again over the endless track, through clouds of dust, often the heads of the leaders could not be seen, and then the jolting was beyond description. We saw many trains of camels, or you might call them strings of camels, they are connected by cords fastened to a bit of wood which is passed through the nose of the camel, the other end being attached to the tail of the one in front. The country, as far as the eye could reach, was nothing but short scrub and sand.

The scenery as far as Boorabbin was very uninteresting, nothing but scrub, with a blue gum tree here and there. From Boorabbin to Coolgardie the country generally was more picturesque, there were more trees and it was more undulating, but the dust, or rather sand, was intolerable.

Boorabbin to Coolgardie 60 m.

Arrived at Boorabbin at 8 p.m., here we had another dinner, and soon went to bed. The Hotel was built of timber covered with

The half-way house, Boorabin.

Our room was next to the bar — only separated by canvas. Four men were playing at cribbage, just as you'd dozed off you'd hear general argument.

hessian (canvas), very drafty. The Pater and I had a room to ourselves — Bird had a shakedown in the passage — no door to our room, just a bit of hessian hung up, and the same in front of the opening leading from the passage to the bush. Slept like a top.

September 8.

Woolgangee to Coolgardie 44 m.

Bulla Bulling to Coolgardie 18 m.

Sunday morning, up at 5 o'clock, good breakfast, started at 6, met the gold escort, which consisted of a coach drawn by four horses. Next to the driver was a policeman with rifle and revolvers, 2 mounted policemen rode in front of the coach and 2 behind all armed with swords, revolvers and rifles. The gold was in an old portmanteau, it amounts to 6,000 oz worth about £20,000.

Stopped at Bullabulling, took up the landlord — Palmer — who played the cornet into Coolgardie.

Saw some beautiful green and red parakeets flying about in the bush, better scenery, more trees. One of the horses fell down, great excitement, everyone turned out, where the track was particularly heavy the driver pulled up and down we all got and walked

Coolgardie in 1894.

by the side of the track — heavy work it was.

At last, at 8 o'clock we arrived at Coolgardie, after a coachride of 98 miles. Put up at the Royal Hotel, place crowded, the Pater & I slept in the smoking room, filthy but slept well. Bird in the drawing room with 7 others. Rose at 6 o'clock and walked around, hired a buggy and drove to the mine. Unfortunately the manager was absent. Started to drive home at 5 o'clock, and got lost in the bush, the driver of the buggy did not know the right track, it became quite dark and we got out and lighted matches to try and find the way. Eventually we got back to the Hotel at 7.30.

Coolgardie.

September 10.

South Londonderry Mine.

Tuesday. Started at 10 for the mine, again in a two horse buggy, met Hicks, the Manager and went down the mine-shaft 112 feet deep, straight ladders 25ft. each very interesting. Had lunch in Manager's tent. Bird and I got picks and prospected for nuggets — but no luck. Started for home, and lost our way again, but not so badly, reached Coolgardie at 7.00.

September 11. Very cold, awful dust blowing hard.

Wednesday, got into the office and put papers straight — at it all day. Bird made financial representative of the Company — jolly lucky to get it. Have been copying accounts all day. All the miners are very civil — Drinks 1/s each, Two drinks every evening — Bird 1 me 1. Like this place pretty much.

The journey to Cue will be more difficult still — Told at Perth that we should die, awful hardships, all nonsense — none at all. Love to all,

Yours Ever,
Bob.

September 15. Sunday. Walked over the scrub for an hour.

Attended church — Text: "Sufficient for the day is the evil thereof."

Buying water at Coogardie.

Coolgardie
Sept. 24th, 1895
(Tuesday)

September 16. Drove to the "First Find" Bullabulling inspected property. Very cold and windy.

My Dear Emma,

September 17. Cold and windy

I had intended writing a very long descriptive letter, but unfortunately I have been so much occupied that I have been unable to do so; I can only give you a description of my journey to Lake Lefroy which took place last Saturday.

September 19. Inspected machinery at "Bayley's Reward".

At 6.30 a.m. I started in the Buggy and pair, which belonged to the Company, with a Capt. Vaudrey — a mining expert. I rose at five o'clock, had a light breakfast, consisting of three eggs and tea, so I had prepared myself for the drive — Lake Lefroy is situated about 50 miles south east of Coolgardie — through the bush, the road being a rough track, some parts well timbered and most beautiful wild flowers of varied colours growing everywhere. As we drove along, Capt. Vaudrey drew my attention to a snake, lying apparently dead by the side of the track. We pulled up and I went to inspect it with the whip. It was not dead, when I hit it. The reptile raised its head at least eighteen inches and shot out its fangs. I then retired and my companion armed with a stick manfully attacked it, again it raised itself up quite 2 feet, but he knocked it over and I hung it over a branch of a tree intending to call for it on our return journey. It measured nearly five feet.

Lake Lefroy [Widgiemooltha].

September 20. Drove to the South Londonderry mine. Stopped at the "Londonderry Int" with Manager, inspected gold. Bobby went down shaft 100ft deep, bad ladders, main shaft 190ft, drive 100ft.

Our first stoppage was 20 miles from Coolgardie, where a reservoir had been formed, about 60ft by 40ft, at the base of some granite rocks — called a blow. As we neared this place we passed a small tribe of Aborigines, making for the water.

After watering the horses, we lit a fire and boiled the "billy", that is the tea (it is always spoken of as "billy" not tea), and partook of tinned tongues, ditto herring and bread. Later some of the Aborigines squatted down not far from our fire, and one of the women who carried a bit of smouldering wood (they seem always to carry this about with them) lit a fire on her own account. They tried to hold a conversation, but it was not very successful. We gave the children some bread and what was left of the other food. They appeared to be perfectly harmless and showed no desire to eat us — after resting the horses for an hour, at 10.45 we again started on our journey. To the next water it was 31 miles, the condenser on Lake Lefroy. The water drawn from the bed of the lake being perfectly salt, it is necessary to condense it. This part of the journey was in parts very picturesque, but at times the track was most difficult, sometimes we had to drive down at an angle of 45 degrees into a deep gully, and then up again. In other parts it was nothing but big stones, any amount of stumps of trees, and deep ruts, and now and then the dust was awful, but fortunately it was a delightfully cool day — no sun.

September 22. Sunday. Had invited Mr Tom Brown, manager of Bayley's Reward to dinner. Being away Bobby had to entertain him. Brown was the son of the author of "Robbery under Arms". "The Miners Right" &c, and who writes under the name of Rolf Boldrewood.

At one point the track diverged to the right, and to the left. Which to take we did not know, however we determined on the left and soon came in sight of the lake, and could see islands on it covered with trees and shrubs, but when we arrived at the edge of the lake it was waterless, but perfectly white from the salt. It looked as if a snow storm had just taken place. I should mention that we knew nothing of the road, my companion never having been in the district before. We kept however, as well as we could to the track, but when we found ourselves on the lake we were a little doubtful as to whether we were on the right route for the mine we were going to see. As luck would have it, in the distance we saw a man crossing the lake (at this point the lake was about nine miles across), it gladdened our eyes. When we met he informed us that he was looking for two camels which had been lost, he then joined us in some whisky and water, and put us on the right track, which continued right across the lake, and on the opposite shore we found the condenser. The horses more delighted than we were to reach the water, they drank six gallons each, at 6d a gallon. The distance we had then driven was 51 miles.

After leaving the condenser we had five miles to drive to get to the Imperial mine. The manager of this mine, Capt. Hoskins, was an old school fellow of Capt. Vaudrey's so he knew we should be well received — it was here we proposed to pass the night. Capt. Hoskins was delighted to see us, he immediately prepared the "billy" and we made a good supper of tinned herring and tongues, and wound up with tinned peaches. It was five o'clock when we arrived.

The mine we were bound for was the

"Perseverance", but it was some distance off, and we had no idea of the way to get to it. At 8.30 I was glad to turn in. This was my first night in the bush & I was about tired out, it had been a long and fatiguing day. My sleeping chamber was a tent about six feet by five feet composed of the thinnest canvas, however, I had three rugs, my overcoat, and Bird's waterproof, so I made myself comfortable, and went to sleep right away.

September 22. Sunday.

It is said the birds do not sing in Australia!!

I woke up at 11.30, thinking it was the morning. After having a look round, I turned in again, and slept until four o'clock. At 5 I got up, the birds were singing beautifully, they only seem to sing in the early morning — during the whole of the drive through the bush, I never heard a note. I then walked down to the lake, about a quarter of a mile off — and examined some machinery being erected in a mine — which I afterwards ascertained was the Cardiff Castle Extended. At 6.30 I heard some one calling, it was Capt. Vaudrey, he thought I might have been lost in the bush. Breakfast was ready, grilled bacon and "billy", very good indeed.

The Cardiff Castle Extended. Machinery by Fraser & Chalmers, timber standards, the ends being cased with lead, a preventative against the white ant.

The horses having been put in the buggy, we started at 7 o'clock for the Perseverance Mine, Capt. Hoskins leading the way in his one horse buggy. We arrived at the mine about 8.30, being Sunday morning everyone in the camp was asleep. We woke up the Capt., who soon turned out with some of his men. The horses were taken out, and very soon some excellent coffee was made, which I thoroughly enjoyed. We then inspected the Mine, and walked over the property which was on each side of a very

The Perseverance, now "Bass & Flinders". I have a specimen from this mine of "tourmaline" carrying gold. No such specimen exists in West Australia, South Kensington, or at Jermyn St.

high hill, from the top of which we had a most magnificent view of the Lake, and the surrounding country. The lake is about 40 miles long, by about 9 to 10 miles broad.

I gave Capt. Vaudrey 200 guineas to report on this property & he condemned it. A considerable sum was subsequently expended on it by a company, which in 1899 was wound up.

At 11.30 we partook of an early dinner — tinned soup, steak etc and coffee — very good. The horses being then ready, we started for the Condenser on the Lake, arriving there about 12.30 — Capt. Hoskins, who had returned to his mine after showing us the way, joined us, it being his intention to drive into Coolgardie, and lucky for us he did so, as you will see. The drive back at first was very hot, but after we had done about 15 miles it came on to rain heavily. This threw us out entirely, we had arranged to camp for the night at the reservoir before mentioned, and to do the remaining 20 miles in the morning. On the road I saw an Emu cross the track, it ran into the bush, I could see it for a long distance, then it turned round to inspect us. I am told they are very inquisitive animals.

When we arrived at the water we were wet through; that is to say, my companion was, he had no overcoat. It was no good thinking of camping out, the ground was thoroughly soaked, so we drove on. The difficulty was, the darkness that had been coming on, and in that country it comes on very rapidly, there being scarcely any twilight.

It was five o'clock when we reached the reservoir, after taking some refreshment and watering the horses, we drove on until Capt. Vaudrey, who had the reins, could see no longer. It was then that Capt. Hoskins came in; he knew the track, so took the

A Goldfields condenser.

lead, driving our horses, Capt. Vaudrey driving his horse. We kept on until seven o'clock, by which time we had arrived at a condenser seven miles from Coolgardie. We, more than wet and cold, and the horses about tired out. Capt. Hoskins made a fire — you can always have a fire in the bush in 5 minutes — and we got out the tinned tucker and had our supper. I wanted it and enjoyed it. The horses had some fodder — good little horses — I drive them every day. Then we started out again — awfully dark, lost the track twice, but eventually landed safely at the Hotel at ten o'clock, had one whisky and went straight to bed. Bobby was at the club, we are honorary members now. I never heard him come in, but woke in the middle of the night, my feet being still cold, but in the morning I was up at seven — well and hearty. Never felt better in my life — Thank God for that and now adieu, with love to you, Lulu and all friends.

Your affectionate Husband
Robert Emeric Tyler.

Returning we saw an interesting sight — a burning tree.

Coolgardie,
Sunday, September 29 1895.

September 24. Went for a drive, Bobby drove the pair. Vaudrey cabled, could not recommend purchase of "Perseverance" mine. Interview with Pell the livery stable keeper. Had charged 20 pound for a buggy, used by Shand the clerk, when he ran away with a girl declined to pay it. The girl's father followed in a buggy and brought her back to Coolgardie.

My Dear Emma,

I hope you received the descriptive report of my journey to Lake Lefroy. I forgot to mention the mirage I saw on the lake — it has caused, so I was told, the death of many poor fellow — the appearance of rippling water was perfect, even to the reflection of the trees and shrubs in the water. Although I knew it was not real, it was difficult to believe it could not be. It was certainly a wonderful sight and most interesting. If only Moses could come once more on earth to touch the rocks in this waterless country so that water might run in dry places, what a happy land it might be. It is this want of water that keeps it back. The price has been as high as two shillings and sixpence a gallon, and even now at a place I drove to on Thursday, it was nine pence a gallon.

September 25. Drove with Hicks to inspect the Challenge mine. One mile from Coolgardie — no good. Introduced by W.F. Arnold.

The manager and I started at 7.45 on Thursday the 3rd to see a mine near Dunsville — 35 miles from Coolgardie — it was 18 miles to the first condenser, the way being through forests of trees, some of the country being very fine, and in parts hilly. One hill is called Mount Burgess, it is of considerable height and entirely covered with trees. There is a gold mine on this mount called after it, and the mine and Mount are named after the man who discovered the former.

September 26. Drive to Dunsville about 35 miles north of Coolgardie.

Water 9d. a gallon.

A Prospector's Camp.

Dunsville.

In August 1894 I.W. Dunn discovered the "Wealth of Nations" and took out £20,000 in gold specimens. In a few weeks if was sold for £150,000.

1906 Dunn was imprisoned at Kalgoorlie for a debt of £20 – prospectors are not a saving class!

From the condenser to Dunsville was 17 miles, but all the water used at that place has to be carted from the condenser. Dunsville was discovered by a man named Dunn. Here is situated the "Wealth of Nations" mine — about 18 inches under the ground they found a 60 oz. nugget. Dunsville consisted of one hessian covered store where everything was sold. The mine we had to see was about two miles from this place.

I should have commenced by saying that we left Coolgardie in rather grand style. We had an outrider who rode in front to show the way, but he was not happy when he arrived at the end of the journey — it was a long ride, nearly 40 miles before we reached the mine, and he had not been used to it.

We arrived on the Mine at 12.15. After inspecting the Mine, and having some tinned sheep's tongues and bread — this should be reversed, we partook of the food first and did the inspection afterwards, but before all we fed the horses. I had christened them Lulu and Kitty, they are a good pair, and go beautifully together, but poor Kitty is weak in one leg, has been over-driven. Lulu is a beauty and would look well in the Park, I wish I could bring her over. You've no idea what a whip I am. I think I even surprised my son, who drove them for a short spell the other day. Well, to return — after the inspection of the mine, we drove to the Miners Camp, which was situated not far from the store at Dunsville, and within sight of the Wealth of Nations mine.

"The Royal Standard."

The chief miner was Mr Slade, the second Mr Hill, and the gentleman who conducted us from Coolgardie Mr Howard. I called him the monied man, because he appeared to be finding the money to assist in working the mine — he was not a digger or delver — he was an elderly man, who had reared eleven children. Mr Slade had a wife, but she was staying in Albany. Mr Hill was single, a big man with a small soft voice. Mr Slade was also a big man but with a big voice.

They treated us well — every luxury, tinned salmon — very good — pilchards, herrings, Irish stew, sheeps' tongues and kidneys, of course, all tinned, and then beetroot and onion salad — not at all bad — washed down with plenty of "billy". After the repast we walked the horses to the store to give them a drink — we all went, to drink also. I had invited them to have whiskies, it was unfortunate possibly that I did so, for after having two each — every drink 1/^{s}. no matter what, lime juice or any thing — Mr Slade would purchase a bottle of whisky at 12/6 to take back to the Camp. I had half a bottle left which we had brought from Coolgardie, but by eleven o'clock it had all disappeared. Mr Slade became loud and boisterous, the soft spoken Mr Hill would constantly rise to sing the praises of the mine — he explained so much about it that we began to think it was not worth touching. When the whisky is in — the wits are out. The monied man who was overcrowded with whisky, at every pause in the conversation said "I've done my duty, I promised to show you everything about the mine and I've done it." He repeated this so often that at last his friends requested him to "shut up". But taking it all together, it was very amusing — they told of their past experiences, related anecdotes, yarned and sang songs, and became quite hilarious.

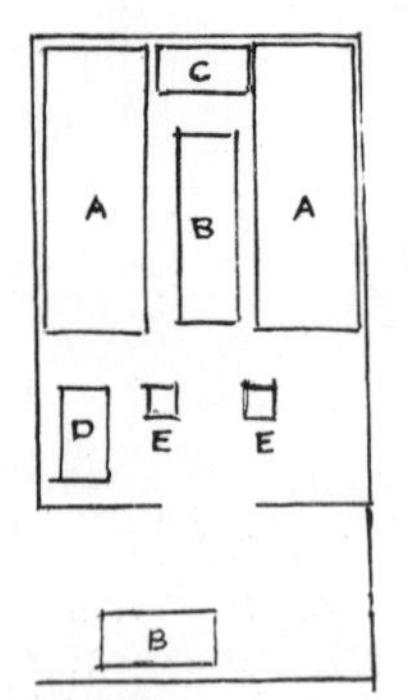

Imagine a tent — a very low one — about 10 feet long by 9 feet wide — AA — shake-downs — B — moveable table — C — a box converted into a sideboard — D — small table — E E — boxes used for seats — this is a correct description of most of the miner's tents.

About eleven o'clock our hosts left us, that is to say staggered out of the tent, their tent being at the back of ours, and exactly the same; this formed their camp. But we had not turned in many minutes before we heard the deep voice of Mr Slade — "I want another whisky" then in a still deep voice "I want another whisky, unless I have another whisky I'll go to the store and have two bottles." His friends tried to calm him, it was no good, so the monied man came into our tent to see if any had been left, but before he could get out again, Mr Slade had departed for the store. We heard that he returned about 1.30 and kicked up a dreadful row, but I slept through it.

September 27.

At 5 o'clock I woke up and found Mr Slade — notwithstanding the whisky — preparing our breakfast. He was cutting up the bacon, the "billy" he had put on, and he apologised for not having any eggs, he'd been all round the camp trying to get some. I enquired how he was. "It's in my head" he replied. Then the quiet Mr Hill turned up. I said — well how are you! Oh, he said, I feel as if I had two heads. I then went in to see Mr Howard, he was sitting up drinking tea looking very bad. Meantime the bacon was being fried on the fire out in the open. With the bacon we had damper made by Mr Slade the night before, it was excellent. With the flour he had mixed ground rice, this he informed me kept it moist — the damper when mixed is put in an iron pan and covered with another, they are then placed in the embers, and left all night; this one was about 10 inches in diameter.

Met Fitzgerald Moore at the Victoria Hotel. Interview with the owner of the "First Find", price about £8,000.

At five o'clock we started, and I drove into Coolgardie in five hours and a quarter — a most beautiful drive, the morning being cool and lovely, so that we could thoroughly enjoy it.

A dingoe's lair had been discovered in the neighbourhood. It contained a collection of old boots & many other very curious articles.

The previous evening I was reminded of Tiny — at the store, a little dog took a fancy to me and followed us back to the camp. As we sat around the open fire, he would get on my lap, just as Tiny used to do, when she could jump. He remained in the tent until we went to bed, then made for the Store. I hope he arrived safely, because if any wild dogs came across him, his sweet young life would be ended, they will always eat a tame dog if they meet one.

September 28.

Drive to Mount Morgan, about 21 miles west of Coolgardie.

Saturday we had another good drive to see a Mine. This time Bobby went with us — We started at 7.20, a lovely morning, like early spring. At first we had to drive very gently on account of poor Kitty's leg, but after a time the lameness passed off, and then she went as well as Lulu. Bobby is very fond of them, and he thinks they now know him. The first 12 miles of our journey was along what might be called the "Queen's Highway", but it is only a track through the bush formed by the teams that come up from Southern Cross, or rather now from Boorabbin, the rails having been laid up to that point.

Seven miles out we stopped at a condenser to give the horses a drink — Kitty drank well, but Lulu turned up her upper lip, and declined to drink. At the 12 mile rocks — that is 12 miles from Coolgardie — we had to turn off to the left, from the main track — a rough plan of the route had been given to us by the owner of the mine we were going to see, and we were directed to follow some wheel tracks, but as there were wheel tracks going in all directions, it was difficult to determine which would lead us to the

Mine. However, we kept on to the left hoping we had selected the right track; it turned out we had, but we pulled up several times to consider the advisability of turning back. The instructions we received were to drive until we crossed the line of the proposed railway, about 3 miles from the main track, then to continue until we saw some buggy tracks bearing to the left — several of them, but showing very faintly — no distance had however been given, but we imagined it would be about 2 miles. Instead of which it must have been at least 4 or more, before we did hit on them. Several times we pulled up, and consulted again as to whether we should not turn back, but we agreed to go on a bit further, and lucky for us we did.

Having discovered the tracks, the instructions were to drive on until we came to an abandoned camp. This, we thought, would be but a very short distance, but it turned out to be another 3 miles, and when we got there we were at a loss to know which way to go, so we all shouted together at the top of our voices, and after a bit we could hear a faint response in the far distance. Gradually it came nearer and nearer, and at last we saw our Miner — the man we expected to meet us. We commenced to abuse him right away. He told us it was about 16 miles and we estimated we had driven at the very least 21. However we were glad for the sake of the horses, as well as our own, that we had discovered the Mine.

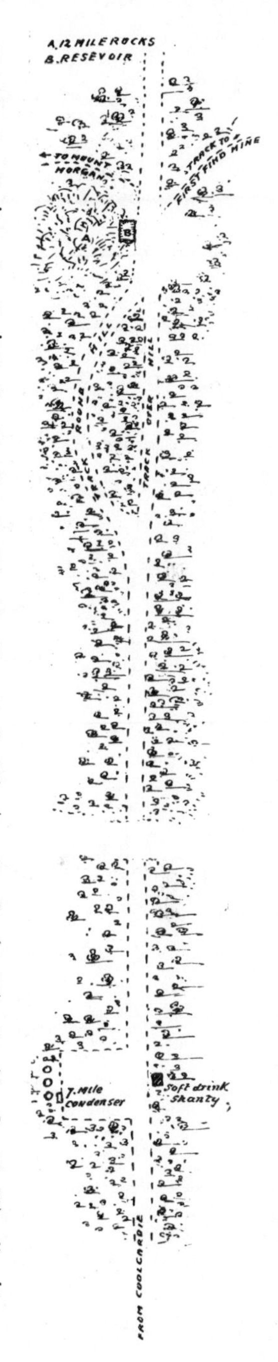

We had tinned sheeps' tongues and bread, and whisky in a flask, lent to me by the owner of the stables where the horses are kept. He had presented us with nearly a bottle of whisky, besides giving us whisky and milk before starting away for the journey. We also had our water bag tied to the side of the buggy — these bags are made of stout hessian (canvas) and keep the water beautifully cool on the hottest day, so you can see that we should have been happy, but not the horses, more especially Lulu, who had refused a drink at the condenser. Immediately on arriving at the camp we took them to the water, which fortunately the owners of the mine had a good supply of. Bobby took Lulu who drank first, then led her back to where we had placed the fodder. He had got her as far as the tree to which she was to be fastened, she looked around for Kitty, but was unable to see her, the water tank being behind a tent, she got so excited, put her ears up and neighed two or three times, until Kitty neighed back again. It was all Bobby could do to hold her, in fact had Kitty not returned her neigh, she would have pulled him back with her to the tank. It was one of the prettiest sights I ever saw.

After this we had a good feed — more sheeps' tongues etc, "billy" and coffee. This was a swell camp, they had a separate tent for the dining room, or Banqueting Hall, only two miners were at home, the one who met us, Mr Walker — the other a Mr Gully, some connection of the present Speaker of the

House of Commons, a young man about 22, had been at the Bedford Grammar School, so Bobby chummed in with him immediately. After having fed well we walked round the property inspecting the lines of reef, and examining any quartz we came across in search of specks of gold.

Gully was the nephew of the Speaker. He died on the Gibraltar Mine in 1902.

Bobby went down the shaft — 45 feet deep, put his foot in a noose at the end of a rope, and was lowered down, where he did a little picking. Then after a cup of "billy" and some bread and jam, we started on our return journey at 4.30. We were anxious to get out of the bush, and on to the main track before dark. Had we been unable to, we should have had considerable difficulty in seeing the track we had to follow. However, we landed safely on the main track just after six, it was then getting rapidly dark, as you know there is but little twilight here, it is quite dark in less than half an hour. The twelve miles drive back was done very slowly, we did not arrive in Coolgardie until 8 o'clock. I left it entirely to Lulu & Kitty to select that part of the track best suited to us, and during the whole of the 12 miles they never touched the stump of a tree, or any other obstacle, and there were plenty on the route. The part we had been to was "Mount Morgan".

I should have mentioned that in parts of the bush which we drove through, grew the most beautiful wild-flowers, all sorts of colours, and some which we picked on our return journey had a lovely perfume. I regret to say, on the road out we committed a shocking and disgraceful murder — I saw a lizard

Wild-flowers.

about six inches long — brown in colour — we all three got out of the buggy and hunted the poor thing until it was dead. I am really ashamed to write this, the poor thing is in a cup by my side. During my former journey I caught a small lizard, about 3 inches long, and got it into a cigarette box. It is now alive and well in the Chemist's shop window next to the hotel, in company with two "devils". These are a sort of lizard about 4 to 5 inches long, with rather round bodies, covered with spikes, most extraordinary looking creatures.

I am now writing on Monday morning, I commenced about seven — I've had my breakfast and it is now 9.15.

Yesterday after lunch at 2.30, Bobby started for the mine, having first removed his Sunday attire and got into his mining clothes. He drove out with Bird (Hercules) and the Manager; he will remain on the mine until probably Wednesday or Thursday, sleeping in the Manager's tent, who will I know look well after him. Bobby likes him, they are quite friends, and he will pick up a deal of useful information from him.

He amuses Bobby, he says "now we must go and find father", and "well! What does father think of it?" The work Bobby is going to do would just suit Percy, I'm sure he would glory in it, fine exercise — he intends to do a shift — which means working for three hours, down at the bottom of the shaft, which is now about 120 feet deep. He proposes writing an account for his University Magazine — "Down a mine by an old boy". It is a great experience for him, he is looking very well, quite fat in the face, but not so brown or rather red, as I am, or more properly speaking — bronzed.

I must say I stand this sort of life wonderfully well. I am always strong and hearty and never off my feed. I wake every morning a little after 5, and am generally out between six and seven. Coolgardie is composed of three main streets running east and west, and two cross streets. The roads are at least 120 feet wide, some of the houses are of wood, some wood covered with hessian. There are a few brick buildings, about six in all. I have marked them B on the plan. All the roofs are corrugated iron, by this you will judge that it is not strictly an imposing looking city.

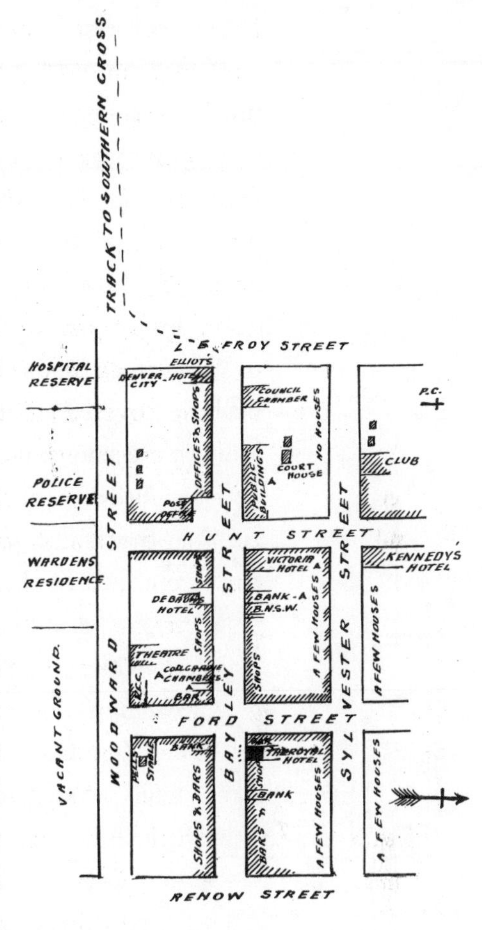

Our Hotel is the Royal, built of timber with usual roof; in front is a wide covered balcony, projecting to the edge of the roadway about 14 feet, supported on timber posts. When the hotel is full, and it mostly is, beds are made up on the balcony. At night, from the balcony, we have a fine view of the "Southern

Cross", these stars are never seen in Europe.

The Royal, as I said before, is built of wood, the walls inside being lined with canvas papered, the ceilings being the same. When the wind blows the whole of the house seems to be on the move.

In the margin is a plan of the hotel — a. dining room, which would seat from 60 to 70; b. public bar; c. smoking room, and the room we first slept in; d.

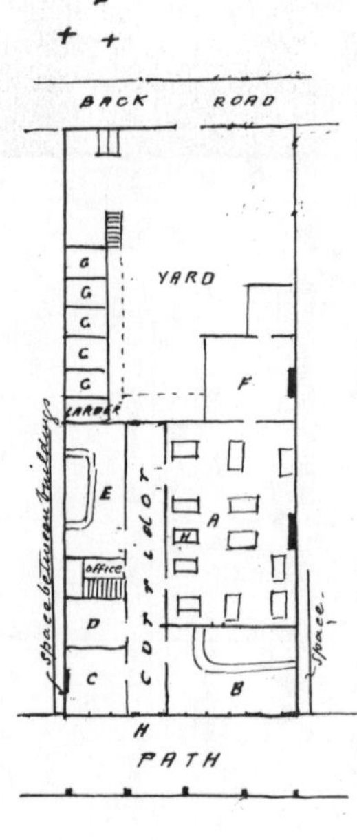

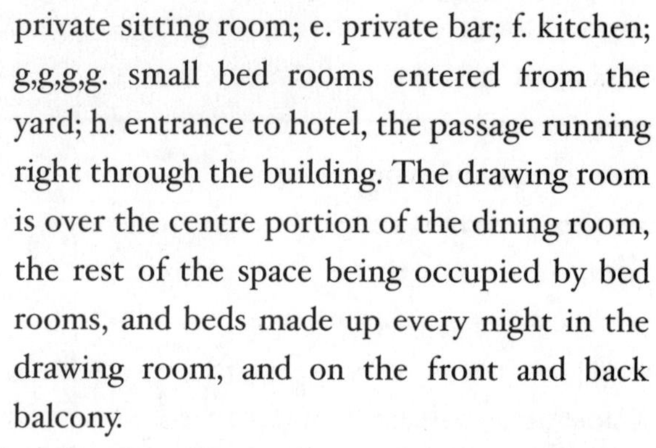

private sitting room; e. private bar; f. kitchen; g,g,g,g. small bed rooms entered from the yard; h. entrance to hotel, the passage running right through the building. The drawing room is over the centre portion of the dining room, the rest of the space being occupied by bed rooms, and beds made up every night in the drawing room, and on the front and back balcony.

The City (they call it a city here, and they have a Mayor) is situated at the bottom of a hill, and when the sun is setting in the West, you see nothing when looking up Bayley Street but 3 or 4 gum trees on the horizon, the effect being very strange. Of course with camels we are very familiar, there being such a number about here. When driving in the bush, if you meet any, and sometimes you come across 30 to 40 or more connected together by a cord, the Afghan who is leading the first camel moves off the track into the bush to allow you to pass. This is the law, as some horses are terrified at the sight and smell of them. In a former letter I mentioned how they were joined together. The man in charge sometimes rides on the first camel, which is always the largest and most powerful. Some will carry as much as 7 to 8 cwt.; occasionally you see some young camels — not bigger than donkeys — running by the side of the mother camel, pretty little things.

You will think it is time I left here. I think so too, and have cabled to ascertain if I am required to go to Cue, as from what I have heard, since I arrived in this place, it may be unnecessary for me to make the trip. I don't want to go, as the coach journey will be much longer and more fatiguing, and nothing to see when I get there, all these mining towns being built on the same lines. Now I guess I've told you about everything up to date which can interest you, but as I am closing this at 11 a.m. Monday, I may before posting it tomorrow, think of something else.

With love to everyone,
I remain,
Your affectionate Husband,
Robert Emeric Tyler

Coolgardie,
October 3rd 1895
(Thursday)

My dear Emma,

In my last letter I told you that Bobby started on Sunday for the Mine to do a turn at practical mining — magnificent exercise, the use of the pick develops the muscles to a fine degree.

October 1. Bobby crushing the specimens from the Royal Standard with our crushing machine. This was the first crushing at Londonderry.

On Tuesday after closing my letters for England, I determined on going to the mine, and as I wanted to leave Kitty — who I mentioned was lame, to see if she would get over it — I determined to ride Lulu, and let Bird drive Kitty. Now when Percy reads this, I fear he will begin to doubt the veracity of his Uncle. You know I have not ridden for some years. I think the last time was at poor Mr Cutt's farm, but I am a bold man, and to a certain extent fearless. At two o'clock we started, at first I must say I felt a bit doubtful as to my riding powers, but in a short time, I fell — no not on the ground — into the way of it, and ambled along in the most delightful style, quite à la Rotten Row. For an hour this would have been most delightful, but as you will soon see I was much longer in the saddle.

In the bush, one of the many disadvantages is that there are no signposts, no people to make enquiries of, and all the tracks look exactly like each other. I had always flattered myself that I had a good eye for locality, but here I am utterly non-plussed — absolutely done. People in the habit of frequenting the bush go by the sun, but the Aborigines, so I am told, are exactly like the Carrier or Homer pigeons — as they are now called — tell them to go to any place they have only once been before, they turn round three times, and then make straight for that place. I'm not quite sure about the three times. I have mentioned these few particulars, so that you might be able to understand and appreciate the position, the most unpleasant position, an individual would be placed in, who was not absolutely certain of the track he had to follow. When we started we believed we knew the track perfectly — I must tell you that there are three tracks leading to the Londonderry, so you

Prospectors' Camp. The Credo, Broad Arrow

have a good choice, and you would imagine it hardly probable that any one could lose their way, but unfortunately there is one other track that we had never heard of, that leads somewhere else, where I don't know, possibly to Adelaide or Sydney, and as ill luck would have it, we dropped on that track — it looked just the same, trees and scrub on each side, ruts and stumps, well it looked as much like the others as two peas.

We gaily ambled on, that is, I did, until at last turning in my saddle — I had to lead because directly I fell — not — as I mentioned before — in the rear, Kitty declined to go ahead — and enquired of Bird if we were all right. Oh yes, he replied and off we went again, until we came to a man working in a hole. He was mining, digging a costeen — another word for trench. They cut a long trench about 2 feet wide, expecting to expose a reef, but more often they don't. I called out — is this the road to Londonderry? I understood him to say Yes, perhaps he didn't. However, on we went again until at last, I felt sure we had lost the track, as by the time I had been in the saddle, I knew we ought to have reached the mine. We had a consultation and determined that the only thing to be done was to "hark back". I am sure we had gone 4 or 5 miles out of our way. I was awfully annoyed to tell the truth, I had had more than enough of the saddle, and I'm sure Lulu had had enough of me; the spirit with which she commenced the journey had departed, however eventually we found the right track, and arrived at the mine at 5.30, so I had been in the saddle two hours and a half. I don't intend to run the same risks again, no unknown tracks for me. No, I did not require any soothing cream, but I was stiff in the joints.

October 2. Bobby went down shaft on Rendall's lease — all white quartz.

Bobby roared with laughter when he saw me arrive, of course I put on a spurt, and came up at a good fair trot. He at once said, you'll be like Percy was at Hastings — I was not. But all day yesterday — Wednesday — I was awfully stiff. I bought a "Lloyds" of August the 25th and turned into bed at 8.30 and read all about old England, the land of my birth. It bangs all the countries I know of. Of course, I can't pretend to say that I've seen Australia, this part of the continent is young, consequently rough, but very little polish. However, I am bound to confess, it is not as it was represented to me by those I conversed with on the way out, neither is it as it was pictured in my mind. I anticipated finding it much more savage, if I may apply the term, more drunkenness and dissipation, more bad language. I have seen very little of the two former, and heard but little of the latter. Talking of Miners, I have met many — they are polite, courteous, obliging and most hospitable, the best they have is put before you, and they will not eat themselves until you have finished. Directly they see you, they ask you to have some "billy", that being the standard drink of the miner, whilst on his mine. The consumption of tea must be enormous, and bad tea too. "Mazinari" sold at 2d a lb. in Ceylon. Bar China and Japan, more tea is consumed in this country than any other — I am told — the reason is on account of the bad water. I've no doubt this explains it — water must be boiled for making tea. At breakfast the other morning we were having a discussion about tea, and the way to make it; one fellow, and a very decent fellow too, an Irish man, or one by descent, said, "the only way to make tea is not to put the tea in at all", at which there was a roar of laughter round the table. This tea sells at 1/6 per pound — we know all about it, Bird was on a tea plantation, so we are up in tea. He won't drink it, but I don't think it so bad. The coffee we have is excellent. I take it for breakfast, and after dinner — at 7 a.m. we have a cup of tea and a biscuit and sometimes well buttered toast, brought to us whilst in bed. I prefer the toast, the biscuits I don't care for. During the time Bobby is at the mine, studying the art of mining, the "Bird" roosts in my room, that is to say, occupies his bed. I think I must be a (very) little like G. A. Sala, he generally after starting on a subject, drifts into something quite the opposite, well I've been drifting into tea.

To recommence — I am still at the mine, I arrived stiff and weary, I threw myself from my noble steed, and nearly sat on the ground, my legs refusing to support their burden, then staggering into the Manager's tent, hurled myself into Robert's shakedown. Whisky was immediately administered, under the genial influence of which I rapidly recovered, and after a brief repose, partook of preserved pine apple, bread and "billy". At 8 o'clock we started back for Coolgardie, arriving at 9.30. The moon shone bright o'er the cloudless sky, not a sound was heard, but the gentle murmur of the wind passing through the gum trees. There was an awful stillness, fortunately not broken by the whoop of the hungry Aborigine panting for a prime cut off a well fed Englishman, or by another Capt. Swift, politely requesting us to "put up our hands." No — nothing exciting or eventful took place, Lulu knew she was going home, but she was tired — my weight had told a tale.

Talking of Bush-rangers reminds me that a man named Morgan is at large in this district. He stole — or as they say "lifted" — a horse, or something of that sort, was caught, escaped, caught again, again escaped, and was retaken. This time they determined to hold him fast, no getting away on this occasion. Two constables were specially put on to watch him, one at a time — there being no prison here — he was placed in a room built of timber, lined on the outside

NOTED BUSHRANGER'S DEATH.

IN SERVICE OF THE STATE.

1901.

A sensation has been created in Perth by the discovery that "Major" Pelly, who died some time ago from a dose of poison accidentally self-administered, and who, up to the time of his decease, was a member of the Civil Service of Western Australia, and acted as secretary to the Government geologist, was the notorious bushranger, Gordon, the original of "Captain Starlight," in Rolf Boldrewood's "Robbery under Arms." In Perth it was recognised he was a man of no mean attainments, and a particularly eloquent extempore speaker. Generally he was reserved, and only on rare occasions was he known to speak of himself. All the information that he ever volunteered was that he had seen active service, and in proof of this he displayed numerous bullet-wounds. From the large quantities of papers found in his apartments after his death it is clear that "Pelly" was highly educated and of extraordinary abilities.

It has been gleaned that the real name of "Pelly" was Frank Pearson, and that he received a great part of his education at Rome. After a wild life in London he emigrated to Australia, where, in 1864, he joined a gang of bushrangers, of which he speedily became the recognised leader. In spite of his associations Pearson, alias Frank Gordon, or, as he was generally called, "Captain Starlight," did not display the coarser or more brutal instincts of the other outlaws, and finally there was a split in the gang, owing to the leader's opposition to useless bloodshed.

Henceforth the gentleman knight of the road was associated with a young fellow named Rutherford, and the pair carried out a number of skilfully-planned and daring robberies of banks and of gold while in transit from the diggings into the large towns. There was a reward of £1,000 offered for Pearson and £800 for his comrade, when on a Sunday evening they were surrounded by the police while drinking in a public-house. In making their escape "Starlight" was shot through the shoulder. The "Captain" having reached his hiding-place in the mountains, sent Rutherford to a Frenchman, a publican, to obtain money, deposited with him, in order that the two bushrangers might decamp from a district which had become unpleasantly warm. It was late when Rutherford reached the Frenchman's place, and he found him dispensing drinks to customers. Having apprised the landlord of the object of his visit, that worthy whispered to him to "stick up" the men in the bar. The young outlaw turned to do so, when the Frenchman, actuated no doubt by greed, throttled him from behind. In the struggle Rutherford's pistol went off, and he was shot through the brain.

"Starlight" vowed vengeance, and one morning set fire to the Frenchman's house, and stayed in front of the only outlet with a loaded revolver, waiting for his victim. But the screams of a woman and her two children inside changed his purpose, and Pearson rushed into the flames and rescued the three at the peril of his life. In the confusion the Frenchman escaped. Soon afterwards the "Captain" was surrounded by mounted police, and after a fierce struggle, in which he shot one of the constables dead and was himself desperately wounded, he was captured. He was sentenced to death, but this was commuted to penal servitude for life. After serving sixteen years and four months, he was released in 1884.

No one in the West Australian Civil Service had the remotest idea of the "Major's" real antecedents. With the skill and daring characteristic of "Starlight's" career throughout, he played a part and played it well.

with galvanised iron. Overcome with fatigue or whisky, the constables slumbered. When they awoke, Morgan had gone — with his hands he had burrowed under the wall, very simple, and very easy. He then walked up to the Warden's house — the Warden is the Chief Magistrate of the district — and wrote his name in the sand outside — he had forgotten his card case — then, instead of making direct for the bush, as on previous occasions, when he was tracked by "Aborigine Trackers", he quietly walked through the City, thus obliterating his foot marks, and up to the present time has not been recaptured. It is imagined by the Police that he is in hiding down some disused mine shaft, his friends providing him with food. They expect to have him soon, as it is next to impossible that he can get away through the bush out of West Australia. If he took to the bush, and attempted to get to the other side — the Eastern Colonies — he would be certain to die for want of water, it is like going across the Arabian desert only there is scrub, trees and sand instead of sand alone. In fact, the desert may not be quite as bad, as occasionally wells are met with, but here there are none. The Chief

Police Camp, Coolgardie, 1894.

Constable dined and breakfasted at my table in the Hotel, at the time Morgan got away. He was awfully annoyed, did not like it a bit.

I heard afterwards that Morgan got away to the coast, and was drowned.

I was telling Mr W. O. Barnier, the Manager of the Bank of New South Wales, about having lost my way when riding to the Mine. He said I will tell you what you should have done — he's an Irishman — you should just have got off your horse, and sat down and had a pipe, then got into the saddle again, leaving the reins perfectly loose, the horse would have been certain to have taken you home to his stable. Nothing like acquiring useful information.

It is now 2.30 — I have just sent the Bird to fetch Bobby and the Manager from the mine and tomorrow morning he will start with the Manager on a journey which will take two days. It is to the mine I wrote about in my last letter. Since then Bobby and the Manager have dollied some of the ore we brought away from the mine, and find it contains good gold, so the latter is going to make a second inspection so as to satisfy himself as to its genuineness. The machine in which the quartz is dollied, is a small machine with crushers inside, turned by two handles. You put the quartz, or ore, broken up about 1/2 an inch square, in at the top and it comes out into a tray at the bottom as fine as powder. This powder is then panned, that is, put into an iron pan about 1'6" diameter, then mixed with water, and shaken about. The water being then poured off. This is repeated several times, until nothing is left in the pan but the gold, if there is any, more often there's none. Bobby's

First crushing on the Londonderry Gold Field.

work was principally to turn the handle, hard work too, whilst the Manager did the panning off part of the business, but no doubt he will give you a full description of this, and what he has done on the mine, when he writes.

Coolgardie Chambers

FIRST Floor

This is a plan of Coolgardie Chambers — our office being marked A, is on the first floor — the only light comes from a sky-light in the roof, but it has the advantage of being cool. The size is 12'0" x 12'0".

I am writing here with the door open so that I see every one who comes up the staircase. In the evening the large hall on the ground floor is used as an open Stock Exchange, and the row that goes on is terrific. One man who sells shares by Auction has a voice which makes you nearly ill to listen to it. I feel inclined to throw the ink pot at him from the balcony, but you must keep cool in this country. This they call the first day of Summer. It is hot — but not uncomfortably so.

October 3. First day of Summer.

I am pleased to say that at the Club, ice is provided, which is a great luxury. I generally drop in

Coolgardie Chambers.

between 4 & 5 and again in the evening. And generally meet someone to chat with, but they one and all have but one idea — that is mining. Nothing but that is the one topic of conversation, the one thought which permeated every soul you come in contact with, from boys who sell the *Coolgardie Miner*, to Mr Shaw, the Mayor.

Mr J. Shaw the Mayor [first mayor of Coolgardie].

I have been introduced to the Mayor, at that particular moment he was lounging on the door step of one of the Hotels. He honoured me by taking a drink, and was quite prepared to take another, and many others. He invited me to call upon him at the Council Chamber, where he said he would be happy to give me any information in his power about Coolgardie, or any other place in the district.

Last evening, about 6 p.m., Bobby arrived from the Mine. He looked thoroughly well, and had enjoyed his mining experiences. He had an excellent dinner, went to bed and slept well.

October 4.

All this morning (Friday) he has been working in the office up to 1 p.m. We then had lunch, and at 2, he started with the Manager to drive to Dunsville, where I drove a few days back. 37 miles, he will be there about 7, and then have a good Miner's feed, and sleep in the tent where I slept. He is certain to be well treated. I met Mr Slade — the head miner — the gentleman with the big voice — today, he was then going to drive out to the mine. Everything he said was ready for them — I am pleased to say he has not touched whisky since I last saw him, he would only take lemonade and claret. Tomorrow morning they will go down the shaft on the mine "The Royal Standard", to get some more specimens, these they will put into bags & will never leave those bags until they are in the buggy. Then they will drive back to Coolgardie. I obtained various specimens when I was there, which, as I said before, having been tested turned out so well, that we want to be absolutely certain that our friendly miners, notwithstanding the large quantity of whisky they put away, did not after we turned in, substitute other specimens, or drop some gold dust into the bags. It is shocking to imagine that any one could be guilty of such duplicity and rascality, but it is done, and has to be guarded against. The Auction has just now commenced — 7.30 p.m. This is what goes on — "Bayley's Reward — Buyers, Sellers, any buyers, any sellers, any bidders" and so the Auctioneer goes on — with a voice which makes your blood boil.

This is termed "salting a mine".

October 5.

Saturday — 9 a.m. Have just seen the gold escort off, an interesting sight — it consists of a spring wagon drawn by five horses — 3 as leaders. Two mounted police in front and two behind, the Sergeant on the box with the driver, all armed with swords and carbines. They go from bank to bank, the specie being brought out in small wooden boxes about 12 inches square. Directly they stop at a Bank, the mounted men in front wheel round and face the wagon. I shall have a photo of it, taken whilst waiting at a Bank opposite our Hotel.

Bobby returned at 7 last evening, and had had a fine time, he will write you all about it.

Gold escort, Coolgardie.

Sunday — instead of going to Church in the morning, we took a long walk into the bush, which did us good. We thought we might have picked up a slug — a small piece of gold — no such luck. When I wrote these last lines, I imagined I should have nothing further of interest to describe, but we never know what the future may bring forth.

In the afternoon whilst we were writing in the office, Mr Barnier, the manager of our Bank, called to ask if I would drive with him to Hannans, a new mining town about 25 miles distant, discovered by a man named Hannan. I should say the first mine in that district was discovered by him — it is now called Kalgoorlie, an Indian name — the meaning of which I don't know. We were to have left at 7 o'clock, but the moon not being up, we delayed our departure until 9. People in England can have no idea what such a drive is like — it is only a track cut through the bush, in parts very bad, deep ruts and stumps of trees, and in some places sand inches deep. This flies all over you, and you get in a frightful state.

On the road we came up with a miner, having all his belongings on his back, walking to Hannans — we gave him a lift, belongings and all at the back of the buggy. He was awfully grateful, we saved him at least 15 miles. Besides which, when we pulled up to refresh ourselves with whisky and water, he came in for a share. He wished us prosperity when we dropped him at the end of the journey — it was then 12.30, every place closed. After putting up the horses at a livery yard, we knocked up the Landlord of the Exchange Hotel, a corrugated iron structure, nearly every building in the place, if not hessian, is of this material.

Kalgoorlie (Hannans).

Exchange Hotel. Rosenthal Landlord. Rosenthal was poisoned in 1906, took poison instead of physic.

When the landlord did appear, he had no empty beds in the hotel, in fact no beds at all, in all the rooms men were lying on mattresses laid on the floor, all he could do for us was to spread a thin horse rug on the bare boards in the feeding room, and give us an opossum rug to throw over us. No pillow, however I made up one with my bag and my overcoat and we laydown side by side. It was hard, awfully hard, I was tired and soon dropped off to sleep, but I woke up later on and found it very cold, the door of the room leading to the back yard being open, and a dog was munching a bone just by me. Further up the room one of the men sleeping on the floor was snoring frightfully. It was no good however turning out, so I turned over and gave my other side a taste

Kalgoorlie discovered by Pat Hannan, June 1893.

of the boards, and soon slept again. At 5 a.m. I woke again, and then did turn out to examine the place in the early morn.

It was a beautiful cool morning, you wanted your overcoat, it was cold enough for that. On looking round I found I was not the earliest up. I should think some of those I saw must have slept standing up, or lying down in odd corners. The fact is, Hannans has increased in population so greatly during the last 3 months, that there is not sufficient sleeping accommodation.

Another hotel is being started, to cost £6,000. Those erecting it have to pay 6 pound per week for the use of the land and only have an eight years lease, that is, at the expiration of the term the Hotel will become the property of the owner of the land. Not a bad investment.

After wandering up and down the one street at present partially built on, I ran across a man we called the Spaniard, who had travelled over with Bobby. He had written me about a mine, we had a chat and an early whisky. I have not taken to drink, but if you don't conform to the custom of the country you are nowhere.

At 8 we breakfasted, liver and bacon and hot rolls, very good, and at 8.20 we started, arriving here at 11.15, a very enjoyable drive, but hot and dusty.

The road for miles, perhaps 10 to 15, is dead straight. You see that line in front of you, and you wonder when you will ever arrive at the end of it. But all roads have an ending — a few parrots flew past, and other small birds. Once I saw a shadow crossing the track, on looking up saw it was a large hawk — a very fine one. We only stayed at the half way house for about six minutes, and had a bottle of beer 4/s.

Everything in the drink line is dear, perhaps this is a good thing for the drinkists. I forgot to mention that at the Hotel I wanted some water to wash in. The man said water is very dear and scarce, so I put off washing until I got here, where it is more plentiful. I then had a magnificent bath, in the wash basin. On a mine, the water used for washing is saved to use for panning the ore, but they don't use much water at any time, even when partaking of whisky they generally take it about half and half.

Well! That's all up to the present. This morning Bobby, Bird and the Manager started at six for the mine, they have a deal of crushing to do, and to do quickly, so they will all be engaged on it. I think Bobby will remain on the mine for the next few days. He seems to like it, he looked wonderfully well after his drive to Dunsville.

Give my love to everybody,

Ever your affectionate Husband

Robert Emeric Tyler

Coolgardie
10th October 1895
(Thursday)

My dear Emma,

I am commencing this letter on Thursday afternoon. No doubt you have seen in the papers accounts of the great fire which occurred here last night. Bobby and I were sitting in the Club talking to a gentleman, and drinking iced lemonade and claret — it was about 9.30 — when a cry of "fire" was heard. We at once rushed out, and saw from the front of the Club an enormous blaze of light. I must tell you that the Club faces a large open space, land yet built on, then comes the new Government offices, not yet quite completed, then Bayley Street, in which the fire was. But it is better to give a plan — what's the good of being an Architect if you don't give a plan!!

Coolgardie fire, Wednesday Oct. 9.

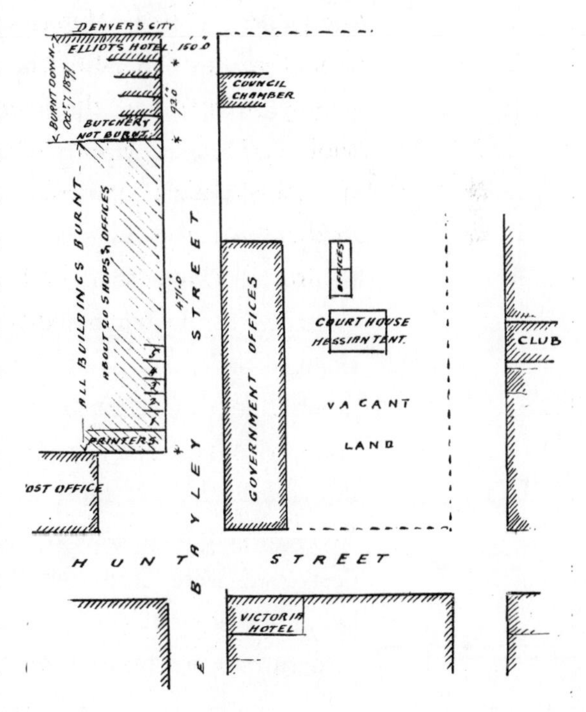

On arriving in

Bayley Street, 1895.

Bayley Street, we found that buildings 1, 2, & 3 were well alight, and as the wind was blowing rather strongly from the East, the shops adjoining were doomed. No. 5 were the offices of our solicitor Mr Harney (Horgan, Moorhead and Harney) who had run out of the Club at the same time as we did — the only thing to be done was to save his papers and furniture. This fortunately was done, Bobby assisting, for which assistance he was very much thanked by Mr Harney. By this time hundreds of the inhabitants of Coolgardie and the outlying districts had arrived on the scene — in fact all but one individual I believe must have been present at the fire. I should mention that the whole of the buildings destroyed were constructed of timber, the roofs being galvanized iron, but at the back of these structures were tents and other buildings entirely covered with hessian. Even some of the doors, or frames forming doors, were covered with the same material, the only brickwork to be seen after the fire were a few chimney stacks, and not many of them. In most instances the fire places are formed of galvanized iron, likewise the flues. By the time, in fact even before, Mr Harney had removed his things, the building was alight.

The Mayor (Mr Shaw) and the Warden (Mr Finnerty) both assumed command, but they disagreed as to the best mode of staying the progress of the flames. However, by this time hatchets and strong ropes had been procured. Excited individuals rushed frantically in to and onto buildings, two or three removed from the burning mass. I should mention that the buildings were only ground floor high.

The first thing done, was to get out of the doomed buildings all the moveable goods and effects. This was carried out by dozens of willing men, the unfortunate owners wringing their hands in despair, whilst others were weeping at their losses. Then the hatchets were freely used, and the ropes fastened to the fronts of the buildings, twenty or more men then laid hold of the rope — I assisted in this operation — and gave a long pull, a strong pull and a pull together, and down came the front. Meantime the men on the roofs were endeavouring to remove the galvanized sheeting, which was really impossible to do in the time, but all worked with a will. In the thickest was Robert, hauling at ropes, removing things out of the shops, and generally assisting with all the energy of a true born Englishman.

All at once a cry of dynamite was raised — some ruffian started the idea that half a ton was stored in the rear of one of the buildings — a general scuttle took place. Some absolutely ran nearly two miles — "they set great store on their lives." I moved, I am free to confess, quickly and so did Robert — the fire however continued to travel on — notwithstanding the efforts made to stay its progress. Then the Warden gave orders to pull down the buildings nearest to the "Denver City Hotel" which is one of the few brick buildings in the City. This was done with a will, hundreds of brawny men, hatchets in hand, demolished the frail structures. Others pulled down anything they could get hold of — two men in their excitement would lay hold of the same thing and pull different ways — had they only known that the wind would shift round, these latter buildings might have been saved, as the flames stopped at the "Buffalo Butchery". The wind had lulled somewhat giving time for the shanty next to the butchery to be pulled down. During the progress of the fire a constant popping, like rifle firing, was taking place. This I afterwards ascertained came from the tinned meat tins, not very dangerous you will say.

The fire commenced as far as one could judge, at No. 1 and went in a westerly direction; it was thought at first, that the corner shop — a stationers & printers — might have been saved, but like the rest it was doomed. It was feared if it once caught light, that the

Post Office would share the fate of the other buildings, so every effort was made to keep the fire back at this point, and all the buildings at the rear of the shop were pulled down; by this means the Post Office was saved, but all the telegraphic instruments, and everything else in the building, was removed. The mail had come in that afternoon unfortunately, and letters and papers were all over the open space in front of the building; these were collected in the morning, but no doubt many must have been lost. Fortunately no accidents or loss of life occurred, only a few fowls, who would fly back to their roost, were "roasted" — bad joke. One horse wanted to return to his stable, it was only the application of a chunk of wood applied to his nose that kept him from being burnt.

But it was a grand sight, one great blaze for about 100 yards — I don't think it had been got up for my special benefit, but I would not have missed it for a good deal. Several people came up to me and said — well Mr Tyler, you'll have something to tell when you get home — you see I am known to nearly every one here, but I don't know everybody. A stranger is a marked man, they all think to themselves, here's another, we can plant a "wild cat" onto. A "wild cat" is a mine that is no good.

At 12 o'clock we had had enough of it, so we made for our hotel. I said to Bobby — where's Bird?, so went to his room and found him asleep — the only man in Coolgardie who had slept through it. We made him get up and look at it from the balcony of the hotel — I believe he regretted being asleep as he

Bayley Street, Coolgardie — aftermath of the fire, 6 October 1897.

missed a grand impressive sight. After we left, special constables were sworn in to protect the goods which had been deposited in the centre of the roadway; you never saw such a medley, boots and shoes, grocery of every description, fancy goods, pianos, clothing, chemist's drugs, dentist's chairs, furniture, bedding, lamps, books, beds etc.

October 10. At six o'clock the next morning I walked up to the scene of the fire. It was still burning in various places, but very little was standing. Any quantity of corrugated iron twisted into all sorts of fantastic shapes was lying about, and two very fine printing machines, they looked as if they had not been much injured. The government is being blamed for not providing a proper water supply. There is practically no water here, consequently no fire brigade — and it is expected as the Summer comes on, there will be another water famine. I trust I shall not be here to experience it.

By the paper sent you, you will read that our late Manager has been committed for trial for embezzlement & appropriating the monies of the Company. This has been a most unpleasant business for me, I tried all I could to get him to make a clean breast of it, but like a fool he would not, although twice he promised our solicitor — Mr Harney — he would, hence the result. I have had another unpleasant experience. The Manager had purchased a crushing machine and paid for it with a cheque of the Company's — about 45 pound. He told Mr Harney he had purchased it for the Company, told him where it was, and that we had better remove it. This I did, sending it to the mine. A man named Pearce — a ruffianly thief — said he had bought it jointly with our manager, and made "a nasty Police case of it" — accused me of stealing it. I had to give it up. The law here is defective, for the reason that the Warden knows but little done about it.

Case against Lowden adjourned until tomorrow. Interview with the Warden at the Club, would rather the case were settled, as it would do no good to the "fields". Would make English investors nervous at entrusting money to people for working mines. Interview with Mr. Hooley (Dalgety and Company).

Charged with stealing a crushing machine.

It would have amused anyone in England to have seen the Court house — a canvas, or hessian, tent, about 30 feet by 20 feet — one door — no window — simply an opening at the back where the Warden sat, and the floor trodden earth. The table at which the Warden dispensed justice — or rather injustice — was covered with a horsecloth, and the reporter had a piece of floor board on his lap to serve as a table. In fact it is about as primitive as it well could be, and yet the inhabitants are howling for "Separation". They want a Governor, and a Parliament all to themselves.

Bobby's first appearance as a witness, and my self as a prosecutor.

Tomorrow I have a special invitation to call upon the Mayor — possibly he wants a subscription for some of the unfortunate individuals burnt out by the fire, or for me to join a committee to do honour to "Michael Davitt" the great Irishman who arrived here on Saturday afternoon, to the sound of the drum, and other wind instruments, but principally the former.

I have just received a letter from Cue, which may entail upon me the necessity of going there to inspect the new mine which has been purchased. But I don't intend to start until I receive instructions from London. The weather is very changeable now, during the day it is awfully hot, but in the evening it becomes quite cool, in fact on Saturday it was so cold towards the evening, that I had to put on my overcoat. The worst of the place is, that the wind blows with great force straight down the street, and the roadway being nothing but sand, it is sometimes frightful, you can't see across the street, in fact sometimes when driving you can't see the horse's heads.

Yesterday — Sunday — we went to church in the morning. It was delightful, beautiful sun shining, with a pleasant breeze, cool and refreshing. After lunch I drove Mr Barnier to Londonderry; it was just lovely, we pulled up at the Londonderry Hotel and had a bottle of beer 4/6d, more than at Coolgardie, but it was real good English beer. Champagne is 25/- a bottle, 15/- a pint, claret and sherry 12/6, and whisky 12/6. Every drink you have — no matter what — one

Price of drinks.

The three men who discovered the Londonderry mine standing. The three men seated in the centre are Lord Fingall, Mr John Finnerty, Warden of Coolgardie, and Mr James Shaw, Mayor of Coolgardie.

shilling. The common miners pay the same price, but they make 4 pound per week, and when on the mine, only spend about 15/-, so they save up and spend it fast when they go into the Town.

In some parts the miners were receiving £4.10.0 a week and two gallons of water a day at 4d a gallon.

Bobby is now becoming a first class whip, he drives Lulu nearly every day, and likes it. If I have to go to Cue, I shall leave him at the mine, he will be all right there, and will have plenty of exercise, which will make him strong and hearty. He can go prospecting and may drop across a good reef.

I saw three men talking in the street the other morning, all three gesticulating. As I came up, I heard one say "Je trouvais un reef" so you see even

Frenchmen are here looking for golden reefs. Many are discovered, but very few are up to much.

Yesterday afternoon Bobby made his first appearance on Lulu, he went out for a two hours ride, and is today suffering from the effects of the same.

We were delighted to hear that you, Lulu and all dear and kind friends and relations were well and happy.

Ever your affectionate Husband,
Robert Emeric Tyler

Dryblowing at Londonderry.

Coolgardie
October 24, 1895
(Thursday)

My dear Emma,

I am so sorry being unable to send you a descriptive letter last week, but only returned on Monday night — as you will read further on in this letter — from a journey, and Tuesday being mail day, and having been away for some days, I had so much to do, that I had no time to write.

I think I will just commence by telling you something about the weather, which I fancy will surprise you. Today has been the hottest we've had, 146° in the sun, 110° in the shade, and whilst I am writing — 8 p.m. — in the office with two doors open, also the skylight, the temperature is 94°. This is a positive fact. We have given up coats and waistcoats, and walk about, and take our meals with our shirt sleeves turned up. Fortunately it being a dry heat, it is not dangerous. This morning I had a chat with a Dr Davy who has rooms just by us; he explained it all, it is a dry heat, and not a damp heat, consequently does you no harm. But it is awful, it makes one so dry, and thirsty, that you are obliged to take long drinks all day long, unfortunately there is only one ice machine in the place, and that breaks down every other day. Lemonade and claret is our favourite drink, or lime juice. Oh for a long lemon squash! My mouth waters at the very thought of it. Fortunately you can get cool water, they hang up the canvas bags (some of them 2'6" long by 8' diameter) in a draught, and the water gets as if it had been iced. But naturally it is unwise to take too much of it, unless adulterated with whisky. We are careful you may be sure. About 10.30 this morning the Doctor brought us two glasses of Claret, with water from his "bag", it was magnificent — most refreshing and cooling.

The following is the description of a journey we made, a short account of which, I think Bobby gave you in his last letter.

On the 17th (Thursday) Bobby, the Manager (not the defaulting one), and I, left here in the buggy at 6.15 a.m., for Hannans (Kalgoorlie), 25 miles. We drove Lulu, and before starting we had two eggs and tea which you will say was wise. Without exception it is the very worst road in the whole district. You become covered from head to foot with red sand, the deep ruts are filled with it, and the wheels going round stir it up, so that after a short time you have the appearance of the noble "red man". The only place to procure water on this road is 11½ miles from here, it is called the half-way house. We, of course, had attached to the side of the buggy a water bag, and we also carried a bottle of whisky, so that on the road, under the shade — and very little of that — of a gum tree, we refreshed ourselves.

We arrived at Hannans about 10 o'clock, then had some bread and cheese and beer, had another horse

Hannan's Street, Kalgoorlie, 1897.

put in the buggy, and drove six miles further to inspect a mine called "Pride of Hills South", owned by Capt. Leyland. Returning we stopped at the "Great Boulder" mine, the largest and most important in the Hannans district. On the payment of 5/- each, we were allowed to descend into the bowels of the earth, but before commencing to do so, we divested ourselves of our coats, then went down, by ladders 150 foot, not dangerous or difficult, as the ladders were only in about 20 feet lengths, that is to say, a 20 feet ladder, then a platform to alight on, then another ladder, and so on. Arriving at the bottom, we inspected the lower workings, which were somewhat damp, then we ascended to the 100ft. level and walked a distance of 1400 feet. In some places we had to crawl on our hands and knees, and over heaps of ore which had been got out by the Miners, ready for removal. It was rough work, and

December 1899, shaft 1100 feet.

my shirt — a white one — was in a bad state when I came out of the mine. We were a party of six, conducted by one of the underground managers; he explained everything as we went on, the character of the stone (ore), and the country we passed through. This is not a quartz reef, it is what is called a dyke formation. A sort of iron-stone mixed up with stone and quartz, but carrying good gold.

As we walked along the drive, the manager lowered himself into a hole, about 8 feet deep. Your son was also down it like a shot, and between them they knocked off a specimen which we brought away, and which shows fine gold. I got on magnificently, climbing up and down shafts connecting the upper and lower level, and as I said before, crawling through holes. But when it came to wanting me to go down a "winge" which is a sort of shaft about 25 feet deep, without any ladder, the only means of getting down being to stretch your legs from side to side, resting your feet on the rough timbers which enclosed the sides of the shaft, I stoutly declined, preferring to crawl back the way I had come. Bobby did it like a miner; I emulated the gentle rabbit. Returning up the shaft is really more fatiguing than going down, and on arriving at the top, I was not only out of breath, but panting for a drink. Fortunately in the office they had a water bag hanging up, so I had a good cool drink.

A winge connects the upper and lower level.

We returned to Hannans in time for dinner. I described this place in a former letter (a beastly hole), fortunately this time I had not to sleep on the floor — arriving there early in the morning, we were enabled to secure a small room opening on to the yard with two beds in it — the size of the room was about 9 feet by 9 feet.

We went to bed at 8.30, tired out, but at 10.30 I was obliged to get up and go to the bar to get a drink. The dry heat is dreadful, your mouth becomes perfectly parched, and you must have something wet. All I could obtain was claret and water, no soda, lemonade or ginger beer in the town and the beer running short. No water to make anything with, and the greatest difficulty to get even a small quantity to wash in. Talk about baths, I don't believe any man in the district has had a real bath for two years. Well you should have seen the state we were in, but I had better not enlarge on this topic, the red dust penetrates everything — your son's back was a picture, he'd been driving all day without coat or waistcoat.

Water was so scarce at the Bars, the whisky decanter being left on the counter, whilst the small jug holding the water was removed immediately. Half a basin of water had to do for three persons to wash in.

The next morning, on inspecting my shirt, I found it necessary to purchase a new one, a neat thing in blue I selected, and Bobby had a similar one.

When we left Coolgardie it was our intention to return the following morning, so took nothing with us, but receiving certain information about a mine, which I wished to see, I determined to drive to it.

After the mid-day feed, at 2.45 we started for Kanowna, thc indian for thc White Feather — 12 miles, arriving there in good time for dinner. This is really the best place we have been to — smaller than the other towns, cleaner and more like (or would be

Kanowna in 1898. An expectant crowd awaiting news of the "Golden Sickle" nugget find.

were the buildings not built of iron or hessian) an English village, trees growing in the roadway, and all round the town. The Hotel also was certainly the most comfortable. We had a good bedroom opening direct on to the street, and in the evening we brought the chairs outside, and sat enjoying the cool refreshing air. That is the redeeming feature of this waterless country. Cool evenings — you can sleep well. Fortunately, when we arrived they had not run out of soda or lemonade — we thanked the Lord for that, and drank deep refreshing draughts. The feeding too was about the best we have had in the whole district, even care was taken to exclude the flies from the butter and sugar, likewise the milk, but they were something dreadful, worse than at Falmouth by far, & that was bad enough.

Sir John Forrest.

I was introduced in the evening to Alec Forrest, the brother of Sir John Forrest, the Premier, who was on the lookout for properties. Several of us had chairs placed at the side of the road, had long drinks and talked — we listened to wonderful snake stories, and big journeys on camel. One young fellow named Vernon, his father is, or was, a member of Parliament in England, had been 800 miles round the country on a camel, and he had many yarns to relate.

Another owned 400 camels, he'd killed a snake 9'6" [approximately three metres] long just as it was about making a good meal of a baby, and any number of tales, about being out for days without water.

Mr Köhn J.P. killed a 9'6" snake.

The following morning (Saturday) I was up at 5, perfectly lovely, had an early breakfast, and at 7.15 started for the "Hit or Miss", that being the name of a mine 20 miles from Kanowna. The one we were going to inspect, "The Florence" being a few miles beyond it. Lulu declined to drink the water at the "Feather". I was not surprised at it, in appearance it more resembled red pea soup than water. At 8 miles we came to a condenser on a lake (no water in the lake), here she had a good drink. I should have mentioned that we had another man with us named Shand; he was driving a small pair of greys, and most

October 19.

of the time I rode with him in his buggy.

We arrived at the Hit or Miss Hotel about eleven. It was a really beautiful drive, right through the bush. We crossed two salt lakes, saw plenty of gum trees and scrub, but not many flowers. The track was only about 10 feet wide, in fact a single buggy track.

The views show the Lake in 1897 after a heavy fall of rain: when we were there it was a waterless lake, and so was the other lake which we crossed.

At the so-called Hotel — it was a galvanized iron structure consisting of a bar, a store — in which everything was sold — a sleeping room and a lean-to, where the food was served, we had tinned herring and English beer, just as a light lunch, then we drove on about six miles to the mine. It was difficult to find, and for a time we lost our way, but eventually dropped into the right track, found the Mine and examined it. It was not up to much, in fact not worth the trouble we had taken to see it.

The Florence Mine.

Lake Gwynne, Kanowna, after heavy rain.

A shooting party on Lake Gwynne, Kanowna.

During this part of the drive I was with Bobby in our buggy, Shand driving in front to show us the way. He had been there before with some of the owners of the mine. Whilst driving along we saw an Iguana — quite two feet, six inches — under a bush by the side of the track. We got out of the buggy determined to catch him if we could, and after hunting him about, I managed to strike him on the head with the end of the whip, which partially stunned him, then Bobby, seizing him by the tail, deposited him in a box we had at the back of the buggy. He was coming to as we got him in the box, and commenced kicking violently. It was no joke holding a reptile that length, and quite 4 inches broad. Bobby didn't seem to relish it much. The Iguana is of the lizard family, but has a large pouch under the throat. They are descendants of the Iguanodon, the "biggest born" of the Saurian race; they rarely exceed, in the present age, more than 5 feet in length; formerly they attained a length of sixty or seventy feet snout to tail — this is known from organic remains which have been discovered.

It had been our intention, instead of returning from the Hit or Miss to Kanowna, to have returned to Coolgardie through the following mining centres, but owing to there being practically no water on the track for the horses, we had to abandon the idea.
Bardock 9 1/2 m.
Broad Arrow 8m.
Black Flag 9m.
25 Mile 14 1/4 m.
Coolgardie 25 m.

We returned to the Hit or Miss Hotel at 1.30 and lunched off chops and bacon — it was rough. No floor boards, for seats candle boxes, this was the roughest place we had been in. We left at 3 arriving at the "Feather" at 7, the last five miles being driven in the dark. Bobby drove the entire day, he is now an accomplished whip, as far as bush driving goes. During this drive we also saw — which delighted Bobby very much — two Emus which crossed the track and ran into the bush. They were fine birds, quite 5'6" high. We dined and slept well that night.

Sunday morning we drove out about 4 miles to look at two mines, and in the afternoon went over "McAuliffs Reward" mine. We were very much interested in this mine. The machinery being all erected, and Capt. Smith, the manager, treated us well, and gave Bobby two gold specimens. After this we went over another mining property and returned to the hotel at 5.30, dined and slept well.

One of the mines we visited was the "All Nations". In 1898 a 23 oz. nugget was found. It is also stated that the biggest nugget yet found in WA. was unearthed at a depth of five feet at this claim 1308 oz. value 4,970 Pounds.

I should mention that on Sunday morning we went to examine our Iguana, but he had gone. Unfortunately there was no lid to the box; we had, as we thought, tied securely round it, the horse cloth, he managed however to worm himself out. I was not really sorry, it seems such a shame to kill such an

The White Feather Reward, Kanowna.

interesting thing, besides which, we might have been doing an Aborigine out of a good square meal — they are, I understand, particularly partial to them. Now this is a fact — as we were driving, I saw a lizard about 10 inches long, so got out to examine it. It ran up a tree, I went quite close to look at it, it looked at me in a most enquiring manner, then I stroked it on the back, it never attempted to move. Evidently it had confidence in my good intentions, and was not deceived.

October 21. Very hot.

At 8.20 a.m. we left for Hannans without any water in our bag. At the hotel they had only been able to get two gallons to make the tea for breakfast.

The horses had nothing to drink; they were offered some, but it was so filthy they refused it. A manager of a mine told me he had to send 18 miles for water, and had to pay 6d. a gallon.

Mr Yabslay, Manager of the Hodgkinson mine on the Kurnalpie road 20 m. east of the Feather.

The drive back 12 miles was very hot and dusty, and dry work you may guess. We arrived at eleven o'clock, no soft drinks to be had, no beer, and very little water. Whilst standing at the door of the Hotel, a cart drove up with soda water — there was a rush for that cart, bottles were frantically seized upon by thirsty souls, who were panting for long drinks. Hearts were made joyful, and thirst for the time disappeared, but no ice, it was not perfect happiness, it is not ordained that we should have all we desire.

At 4.10 we left Hannans. There was all the appearance of a coming storm, the wind rose with great force, and as we drove on the dusty track, at times, the horses' heads could not be seen. The dust and the heavy road prevented us pushing on rapidly, so it was 7 o'clock before we reached the half-way house. Lulu had 4 gallons of water at 9d. a gallon — drawn from a cask behind the bar. We refreshed ourselves with bottled beer at 4/- a bottle. I gladdened the hearts of several thirsty miners by standing them long "Australian" beer drinks.

The remainder of the journey was made in the dark, but fortunately the moon rose so that it was not so bad, nevertheless, we managed to get on to the wrong track, ran up a heap of mullock, and nearly landed into a dryblower's hole. But all's well that ends well, we arrived at the Royal Hotel safe

At one time when the moon was clouded, we found ourselves going into a shed, a voice called out – "where the devil are you coming to". The voice then directed us to the track for Coolgardie.

and sound, had some eggs and tea, and then went to the Club. Bobby went to bed — he has just denied this, and I now recollect he went with us — Bird and myself.

Our solicitor (Mr Harney), an Irishman, was in an excited condition, shouting (that's the expression here) drinks to everyone in the Club. He is a fine big fellow, with a clean shaven face, and brown curly hair; his costume consisted of a white shirt buttoned at the back of the neck. Well! Of course, he had trousers on, and a belt. A Mr Elburn, the wit of the club, a nice jolly little fellow, not up to Harney's shoulder, tapped him on that part of his body where his waist coat would have been, as if he'd been a telephone, saying "Are you there — You've forgotten me." On which Harney looking all round, and then down, caught sight of him, and said, "I thought this place had been well swept out." The laughter at this was uproarious, and immediately more drinks were passed round. At 12 o'clock we left having had a long and fatiguing day.

The following is a true tale. Many years ago, a cook on board a ship cut the Captain's head off and served it up well cooked with a tasty sauce. He was not hanged, probably he was considered to be insane, but was sent to the Swan River Convict Settlement; he is now a respectable inhabitant of Fremantle, carrying on the business of a tobacconist.

The largest bookseller and stationer in Perth (B. Stein & Co.) was a convict. If you are found walking in the City of Perth after eleven o'clock p.m., the police can demand your name and address in accordance with the old convict law which still exists.

Here they hanged for attempted murder, two men, burglars. Not long since they were discovered by a policeman, he was struck with a jemmy, being injured for life, they both were hanged.

The law here is made for the poor, not for the rich, directly a man becomes wealthy, all hands are against him, the one idea is to ruin him, the only thing he can do is to clear out. A man was a Clerk in the office of the Superintendent of Railways (the railways mostly belong to the Government). The Superintendent discharged him for insolence and drunkenness. He was made a Member of Parliament, got the superintendent discharged and took his berth.

We hear that there is nothing but bribery and corruption in the Colony; this comes of a purely democratic government. An honourable man stands very little chance, in fact he is quite out of the running, consequently the worst class obtain the control of affairs.

About the Mining Expert. They are not held in high esteem, as the following shows; "The Liar, The d—d Liar and the Mining Expert.

When talking this morning of hanging for attempted murder I said, if they were to hang a few of the "Mining Experts" who gave false reports on mines, they would be doing the country some good. I don't suppose out of the hundreds of mining experts, and mining men here, that 5 per cent are honest, trustworthy men.

October 24.

On Thursday afternoon, after the terrific heat of the day, we took a drive before dinner for an hour,

A corner of Toorak, Coolgardie.

just to get us cool for the evening. The track we took we had not driven before. It was a good road and very pretty country, the ground in parts being covered with yellow flowers. Bobby has made the discovery that wherever iron stone is, yellow flowers abound. Returning, we had a slight discussion with the driver of a cart, which delayed us for a few minutes. The driver was on the wrong side. Bobby pulled off the track, but the driver did not; consequently the wheels locked. A strong argument followed. The driver contended that his cart being loaded, he could drive where he liked, and was not obliged to pull out of the way. We held a different opinion, however no damage was done, and nothing passed but angry words.

The Toorak Road. Toorak is about one mile from Coolgardie, and is the swell suburb. Villa – residences – constructed of timber, iron and hessian.

Yesterday, Friday, we drove to the mine about 12

o'clock, taking sandwiches with us. It was a really lovely day, cool and beautiful — a perfect hurricane had been blowing all night, and apparently had blown the heat away.

I went down the mine to see the progress that had been made. We are getting into better country, and our manager fully expects to come across something good within another 50 feet. All is conjecture, but still there are certain indications to go by. Leaders are found in a shaft (leaders are supposed to be the commencement of a quartz reef), they vary from say 2 inches to 6 inches in width. In our case there are two leaders 2" wide bearing fine gold; they are 10 feet apart, and it is supposed at a depth of 100 feet they will junction, and form a good reef. I don't suppose

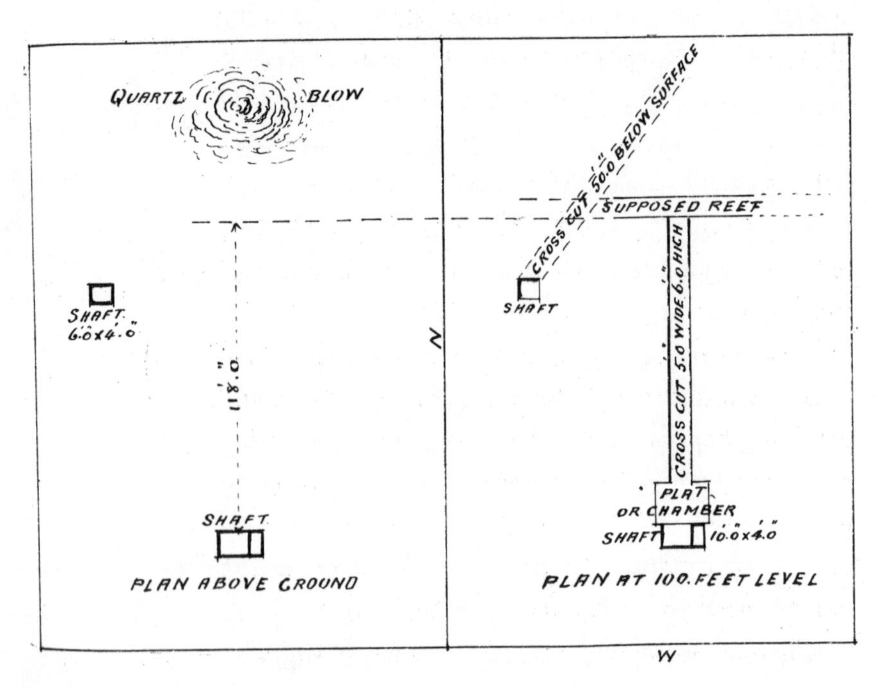

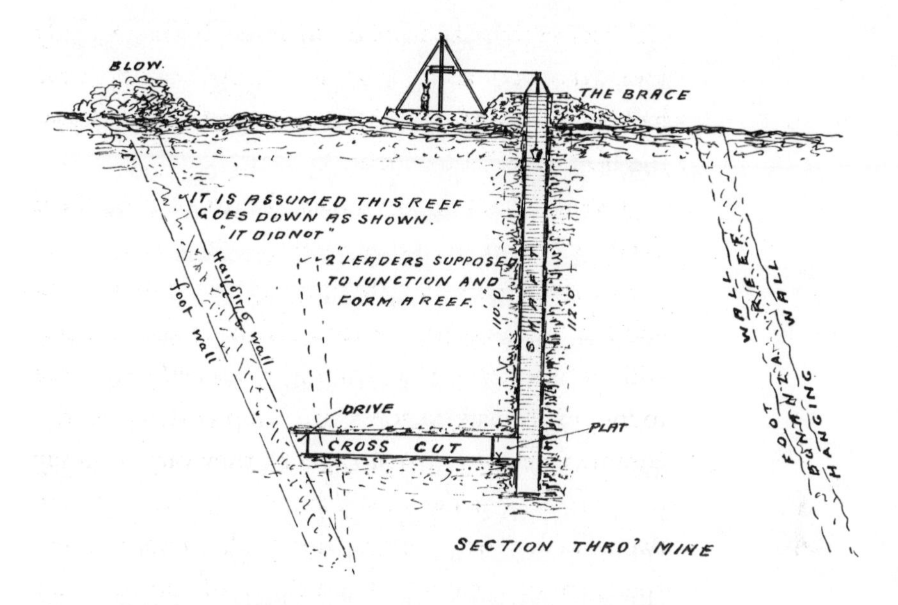

this will be understood but I will explain it when I return. A "blow" means a mass of quartz appearing above the surface of the ground. This particular blow is underlying to the West, so that it may be met with at the end of the cross cut. In fact, it may junction in with the supposed reef formed from the leaders before referred to, if this should be so, and the reef gold bearing, then the mine may turn out better than we have expected.

In a former letter, I mentioned that Kitty had been taken to the mine for a rest, to see if she would get over her lameness. Meantime Lulu has been doing the work alone. As however I had arranged to go with Mr Barnier on Sunday, the only day he can get away, to see a mine some distance off, I determined to bring Kitty with me back to Coolgardie. So we had a strong rope fastened round her neck and then tied to the back of the buggy. This was done without consulting Kitty's view on the matter, and we very soon discovered that her views were strongly opposed to being tied to the back of the buggy. She commenced pulling backwards whilst Lulu was pulling forwards. It was a sort of a tug of war, and we were in the middle with the prospect of being turned out. However, I managed to get down and commenced leading her, not a pleasant thing to do on a heavy dusty road, then Bobby had a turn. Eventually he arranged a seat at the back of the buggy, and by dint of coaxing and talking to her, after this style — "good little mare", "good Kitty", and other such endearing terms — we managed to get her along fairly well. Occasionally she became frightened and tried to pull away, but at last we got her to a canvas tent where they sell cooling drinks, and for a money consideration of 10/- I induced the noble proprietor to walk her to Coolgardie, a distance of about 5 miles. Well pleased we were to get rid of her.

I've seen, just off the principal street of Coolgardie, poor horses standing round a galvanized water tank licking it with their parched tongues. It was a sad sight!

Whilst arranging with the man, we saw a poor grey horse (he was a good horse, but thin) standing by the tent. The poor thing wanted water, but the man said "I gave him water yesterday, but at 5d. a gallon, I can't afford to give him any more." We gave 1/-, and the poor horse had his drink, we saw him have it. Many of these poor horses who have strayed away, die in the bush for want of water and food.

October 26.

This afternoon (Saturday) we drove to see the "Polo" but unfortunately missed it. I am pleased to say that the rest seems to have done Kitty good. She showed no signs of lameness, and I think will do the journey tomorrow all right, and without any difficulty. I drove her today so as to get her into condition for the journey.

The Polo ground is on the Hannans road, we then took the most westerly road to Londonderry. Fine undulating country well timbered, passed by a deep creek, but no water.

Sunday at 9 a.m., we started for our 15 mile drive. It was hot and dusty; you can have no conception of the dust, it hangs about the road like a cloud, fills your eyes, nose and mouth. The only thing you can do is to pull up occasionally and partake of some water out of your bag, mixed with a little whisky. We always take a bottle with us whenever starting on a journey, it might be the means of saving our lives.

The First Find, Bullabulling.

We arrived at the mine at 12 o'clock. Bobby took the horses out, he now knows all about that sort of thing, he's a regular ostler, and after they were watered and fed, we entered the tent where the miners live. They had a tent within a tent, the small one inside was the sleeping apartment, the other space the dining hall. They fed us well — ox tongue tinned, peas cooked in the tin, and excellent they were, good pickled cabbage, and preserved peaches and "billy". We were all hungry after our drive and enjoyed it thoroughly.

Then came the inspection of the Mine. Now don't get excited at what I am about to relate. The shaft was 100 feet deep, but was not constructed with ladders at stages of 20 feet or thereabouts. Simply one long ladder; to one unused to such a ladder, it naturally would be somewhat dangerous. It was all very well for miners. The only other way to descend

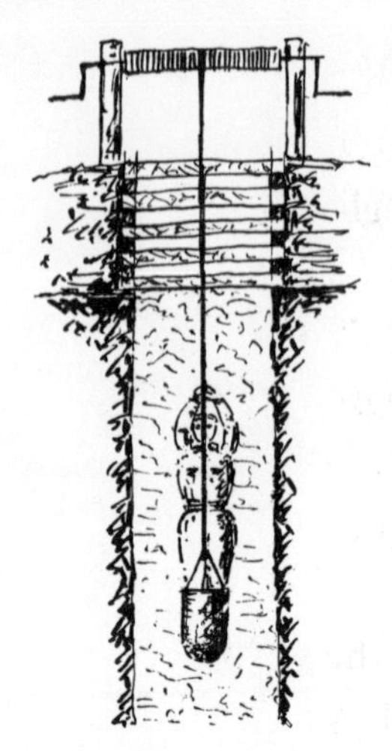

R.E.T. descending the Shaft at the First Field

was to put your foot in a noose at the end of the rope, or stand up in a cow-hide bucket, used for bringing up the stone from the bottom of the shaft. This is the way I went down — side sketch — arriving safely at the bottom, in fact, doing both journeys without accident. I can't, however, say I liked the sensation, to be suspended in the air, with nearly a certainty that if you "lost your head" or the rope gave way, that you would be hurled into eternity, was not pleasing. I hope my brother Directors and the Shareholders generally will appreciate the manifold dangers and difficulties I have passed through. I should have mentioned that the rope was attached to a windlass worked by two men; when suspended in mid-air the bucket kept turning round, like a spit before the fire. To stop this, it became necessary to leave go the rope with one hand, which was not pleasant, and steady myself against the side of the shaft.

I am now writing in the morning — 9 a.m. — two events occurred last evening; our attention was called to a young dingoe (the wild dog of the country) who had entered the building and ascended the staircase. The man who drew our attention to it said he had lived in the country many years, but had never known a dingoe enter a dwelling.

The dingoe.

The dingoe is of a sandy color, foxy about the head, with a somewhat bushy tail and is about as high as a small sheep dog; the bite is very dangerous — event number one. The second might have been more serious. We have a small lamp, I don't like lamps — never did. Something went wrong with it; it commenced blazing up — the oil had taken light inside; fortunately it is of metal — not glass — otherwise we must have had an explosion. We managed to get it out, and I don't think we shall light it again in a hurry. I fancy the great heat must have made the oil hot. I can assure you the water in our bed room is sometimes quite warm.

Mr Rogers, the gentleman who pointed out the dingoe, came to thank me for offering him Bobby's bed. On Sunday morning about 6.30, I found him lying on a sofa on the back balcony of the Hotel; he looked ill, and knowing him slightly, enquired what was the matter. It appeared he had been thrown from his pony at Polo — it had taken place after all — and was badly hurt. He is now in a room in this building, on the same balcony as we are. Had it not been for a good dose of Scotch whisky, he considers he would have been much worse.

There appears to be great efficacy in Scotch whisky (Ushers or Dewars). He was telling us of a Mining engineer who called at his camp, and, partaken of an over dose, he fell down his own shaft, 100 feet deep, broke both arms and otherwise injured himself, but he is now doing well, and will live to partake of more whisky.

I think I have now exhausted all my news; you know everything up to date, we are both well, eat and drink well, and sleep well. What more could be desired. All we want now is to begin moving towards home, but the Company insist upon my remaining until I have found them a first class mine. This I have done in the Mine, the shaft of which I went down, and I hope sincerely they will accept it. I don't want to remain here any longer, I want to come home.

Bobby says that I write so fully that when we return we shall have nothing to tell, but still I know you like to hear everything, and it is a pleasure to you, and every one else interested in our welfare.

With fond love to yourself, Lulu, and all relations and friends,

I remain,
Your affectionate husband,
Robert Emeric Tyler.

Coolgardie, November 4th.

October 28. Frightfully dusty with hot winds, in the evening cold and dusty. Much influenza about.

My dear Emma,

Having written such a long letter last mail, you must be satisfied with a short one this mail, and as a matter of fact nothing of an exciting character has transpired.

I will commence by saying that the weather for the last few days has altered for the better. It is much cooler, more pleasant, and consequently more endurable. Whilst I am writing it is only 78°. Friday evening it was so cold that we had to return to the Hotel and put our waistcoats on, and later in the evening, you could really have put up with an overcoat.

Water is continuing to be awfully scarce, and we are now washing in salt water. At first I could not make it out why the soap would not lather, but I soon discovered when I put it on my face and lips, it was just like sea water. This morning I washed in a tumbler of water out of the drinking bottle. Oh for a bath of fresh water!! I shall revel in it.

The "celebrated" Londonderry mine was promoted by the late Col. North. Bobby went down it. I was satisfied with a view of the rich gold quartz which was kept locked up in safes. Mr Aarons was then managing the mine. They commenced crushing with a 5 stamp battery. Stopped at the Londonderry Hotel and refreshed.

Last Tuesday after posting the letters, I drove Dr Davy to the Mine, and Bobby rode Lulu. A long ride for him — 10 miles there and of course the same back. He did it all but the last three miles; then having a pain, the Doctor finished the ride. He can now drive and ride which is something to say when he returns. I think he will have a camel ride before returning.

November 1.

On Friday we again drove to the First Find Mine, taking the mining expert Capt. William Vaudrey with us. Bobby had to sit behind the buggy. We started at 6.30, a beautiful cool morning, but the red dust that settled on your son was something awful; he again looked the picture of a red Indian, the only thing to do was to scrape it off. I did not go down the shaft again, but Bobby did, and assisted the "expert" in collecting specimens. We dined in the Miners' camp off mutton chops. We did enjoy them, and preserved peaches, billy and whisky and water to follow. I drove the horses the 15 or 16 miles without them turning a hair, over a heavy and dusty road, and brought them back so fresh that they tried to bolt away in Bayley Street; it was down hill, but I managed to pull them up before they got far.

The Londonderry mine was discovered in May 1894, and thousands of ounces dollied out of it. Lord Fingall purchased the claim for 180,000 pounds and an interest.

On Saturday we drove to the Polo ground, but it turned out to be only a practice. I am expecting to leave for Perth on Wednesday.

November 4 96° in the shade. November 5 82° in the office. November 6 Much cooler.

The 6th of November, my birthday. Only imagine at the somewhat advanced age of say 46, to be in this sunny, but waterless land. What a country it might be, had it only water.

As water rises in price so does washing; this is our last bill: 1 suit of pyjamas — 1/6 — 10 handkerchiefs 2/6, 2 undershirts 1/6, 2 socks 8d., 2 collars 6d., 2 soft shirts 2/-. What they now charge for a white shirt I can't say. I suppose about 2/6. Imagining we were only coming here for a fortnight, we left nearly all our thin things at the Hotel at Perth, it being quite cold when we came up, so we have had to purchase different articles to make up. Unfortunately, when you purchase ready made trousers, a thing I've never done before, you find no provision is made for gentlemen with fully developed figures. They assume a man 38" round the waist — my measurement — must necessarily be provided with long legs, the result being I have myself had to turn up the trousers I purchased. I turned them up on the inside and then sewed them in four places; it made a beautiful job, the only fear being that the stitches may come out. I live in hopes they will not.

Here is a sketch of the hat I am wearing, the great advantage being that it keeps the sun off, and is well ventilated, there being a space between that which fits on your head and the hat itself. I have asked Dr Davy and his wife to dine at the same table with us tomorrow night. They always feed at the hotel, and I shall stand the wine etc. They have some good Chablis (Australian) at 12/6 a bottle, and when mixed with a soda water makes an excellent drink. I know you will drink my health and safe return, and I shall drink "my safe and quick return." With fond Love. Believe me your affectionate Husband.

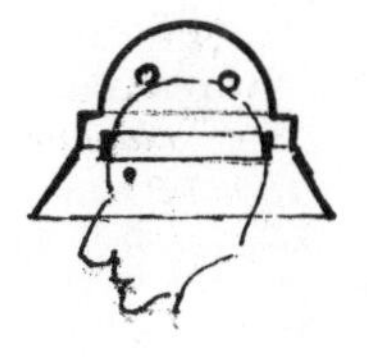

"My hat"

Robert Emeric Tyler.

Grand Hotel, Perth. W.A.
Nov 17th 1895, 5.45 a.m.

My dear Emma,

My last letter gave you a faithful account of our doings up to Tuesday November the 5th, Guy Fawkes Day — not a guy was to be seen, neither did I hear the subject referred to (it is now striking six, and I am sitting on the side of my bed writing this letter).

We were very busy all day on the 6th — had paid the deposit on the "First Find Mine", the new mine I am acquiring on behalf of the Company, it should be a lucky one, and I trust most sincerely it will be, for the shareholders' sakes. It is all luck in mining, no one can tell what is in the bowels of the earth. They can alone judge by what they see, and the general formation of the country. They may see what appears to be a good gold-bearing reef, but it may "cut out" as they term it — that is, instead of continuing through the property, finishes off abruptly. You can but take the opinion of the best, and most trustworthy expert. We have done so in this case, Capt. Vaudrey, he who accompanied me to Lake Lefroy, an old Cornishman, has given an excellent report, in fact, had I not purchased, he would have taken it, at a higher price for a Syndicate he acts for, so I hope I have done good for the Company.

The Lulu mine adjoins the First Find, and may turn out as good a mine. In the evening Dr Davy, his wife and Mr Barnier dined with us. We had two bottles of Australian Chablis, a very good wine indeed, and one bottle of Champagne. They drank my health, and I proposed absent friends. Of course the first was yourself, the second Lulu, then all dear friends and relations. The toasts were responded to with hearty good will, and we talked about you all, which made Bobby and me think of home, and that we were nearing the time we should see you all once more. We dined at 7, the hour we should have dined at home, but the time differing so much — I think we are $7^{1}/_{2}$ hours in advance, so you did not drink my health until much later.

I know, of course, that you drank my health, and I make no doubt many tears were shed on the occasion, but I also hope that you enjoyed yourselves, and that you were surrounded with the members of my family, and yours, and those interested in our well being. I must say that I never felt better, and we were in good spirits because of having settled the purchase of the mine. I had been cabled not to leave Coolgardie until I found a mine suitable for the Company, I had found one, so that I felt I was free to return to the land of my birth, and to those I love so well. But this is a digression; we finished the dinner with Café, liqueurs and cigars, them went to Dr Davy's rooms, and had some whisky, and a general discussion. He is a very clever man, brought up at Oxford, and he can talk; I should imagine he is a good doctor, fortunately I have had no necessity to test his capabilities. I wanted Bobby to consult him after his 20 mile ride, but he declined. Besides doctoring, he understands minerals, and mining, quite as well as most of the experts, having studied at the School of Mines. He used our offices, and examined our specimens, and told Bobby all about them. I can assure you he gave Bobby excellent advice. He is a man about 40, stands 5'9", curly brown hair, blue eyes and has a bright smile. His wife is a pleasant woman, but suffers from a "cock eye" — you know what I mean — you don't always know who she is looking at — they have three nice little children, but they are located about 8 miles from Perth.

After leaving the doctor's, we went to the Club and had a game at Billiards, getting to bed about 12.30. The following day being the day before my departure, we had a deal to do seeing people, and making final arrangements in connection with the two mines, so that we were unable to commence packing before quite 11.30 p.m. We were nearly too tired to pack, but it had to be done, as we had given instructions that we should be called at 3 o'clock. The great difficulty we experienced was to get our things into the portmanteau. They had increased so, owing to our stay being so much longer than we had anticipated. Our old trousers etc., which had been used during our mining excursions, we left behind, even then the portmanteau would not come to.

We turned in at 12.30 hoping that sleep would give us strength for our journey. At three, to the second, we were called, vigour had returned and we got the portmanteau to close within four inches, and it had to go like that. Then we had an egg and some tea, and started for the coach, timed to leave at 4 — I looked after the portmanteau, Bobby after the box of specimens — rugs we had none, they'd been lost at the stables (the value of them will be deducted from Reggie Pell's account, he is the keeper of the stables), but we had no need of them that journey.

November 8. Started for Perth.

At 4.30, we started — 5 horses — full inside and out, we had outside back seats. There were two ladies inside, Mrs Viner and Mrs Saunders — their husbands saw them off. They wept not, neither did they sigh (I refer to the wives, what the husbands did I can't say) — on the contrary they were quite happy and gay in the society of the gentlemen inside. Possibly they were more gay at the thought of leaving the City of Coolgardie — who knows?!

Mrs Viner was the wife of a ruffianly solicitor, mixed up with our defaulting Manager, and in fact defending him. He has had 100 pounds of our money and won't return it. The other was the wife of a clerk in the Warden's office, both somewhat pretty, and both flirts. Mrs Viner was old enough to know better; the other being only about 22, it was excusable — youth will have its fling whether married or single. And from what I hear the ladies of W.A. are much given to flirting, possibly the atmosphere may have something to do with it.

The standard of society is not high here. The brother of the Premier — Alec Forrest, who I met at Kanowna — was blackballed at the Coolgardie Club, he not being looked upon by some (a very few) as a highly respectable member of society, and some will not visit them or their wives. The men inside the coach were Mr Haynes, a solicitor — Mr J. Hare — ditto — Mr Leake M.P. — ditto — and M. Faddy, a mining man. The latter I knew, and a good sort — had been brought up on the "Worcester Training Ship", but having met with an accident had to give up the sea and went in for mining.

But this is all by the way — we are at the back of the coach. I object to the back of the coach, because every jolt — you have one on average every ten minutes — you feel that you must be precipitated head first into the road; and then the dust, the 5 horses and the lumbering wheels stir it up in such clouds, that you are completely enveloped in it. Besides this, you have the sun right in your face, travelling due west in the early morn as we were, we had the full power of its rays. The perspiration was streaming down our faces, the effect being that in a very short time we looked like — as I have before said — the noble red savage of North America. The heat in the morning was not quite so bad, but in the middle of the day it was near roasting point. The first stage was 18 miles to Bullabulling, where we breakfasted — a long run for the horses; the next was at Yerdi — 14 miles — then 12 to Woolgangie, and the last 16 miles to Boorabbin, where we arrived at 2.45. The final stage was too much for the horses, we had to get down and walk the last mile or two, which produced an intolerable thirst, which we had the

The tank at Bullabulling held 1,198,000 gallons.

Southern Cross.

pleasure of slaking directly we arrived. We also had a good wash and general clean up, which we stood much in need of.

At 4 o'clock we started by train for Southern Cross — 61 miles. This occupied 4 hours — 15 miles an hour; the train consisted of two long carriages opening into each other. Ladies, Members of Parliament, Solicitors, Architects, Mining experts and the great unwashed all herded together. No stoppage where water or drink could be obtained for 30 miles — 2 hours. I soon became dry, and seeing a miner opposite to me had a water bag (we are bringing one home with us), I asked him for a drink. He was real good sort, had his arm in a sling, the result of an argument with a Frenchman. He had applied his fist to his nose — that is the Frenchman's nose — who, not appreciating that mode of settling an argument, had stabbed him in the arm; he was going to Perth to spend a week or so with his wife to get over it. Well, like a good fellow he handed over the bag and we all had a drink and most refreshing it was; what would be done without these bags I can't imagine. At the 30 mile stoppage a man sold "Hop Beer", beastly stuff, but it was wet — I was also successful in getting a bottle of water, which was put into the friendly miner's bag; this lasted us until we arrived at Southern Cross.

At 8 o'clock we sat down to dinner, the best we had had for some time. I got the waitress to get me a bottle of water for use during the night journey. I gave her 1/- for waiting on us; she wanted to know what it was for, did not understand receiving tips, had never had one offered before — how different to the old country.

The train left for Perth at 10 o'clock. When we arrived on the platform, we found there were 6 first class carriages, 3 of which had been reserved for Mr Leake M.P. for Albany. Not a bad sort, we had several conversations with him; he sat next to Bobby coming up from Boorabbin. His party consisted of the two ladies, Mr J. Hare, the partner of the ruffian "Viner", and some others — this meant that the 3 remaining carriages would be full up. My indignation rose at this, but being somewhat of a philosopher, it occurred to me that instead of being annoyed, I ought to be thankful that we only had one M.P. with us; had there been two, no doubt the other 3 carriages would also have been reserved, and we should have been left to ride in the goods van, or on the engine, the second class carriages being crammed full of miners who had probably not seen water for washing purposes for months.

It ended with us being in a carriage with five others, so there was no chance of lying down.

At 2 a.m. we stopped at Hines Hill for early breakfast — we first had some whisky and milk, then two plates of roast beef, mashed potatoes and greens and beautiful lettuce — so white — Oh lovely. We were nearing civilisation (thank God). We completed the meal with rock cakes and coffee. After this excellent repast we slept well. At 4.30 I woke up, washed my face and hands & otherwise refreshed myself, the water bottle — that is the water — came in handy — and I viewed the scenery we passed through — some of it very pretty.

At 7.30 we stopped at Northam, good Hotel, good breakfast, and any quantity of beautiful new milk, direct from the cow. Oh civilisation!! what a charm thou hast. We began to feel elated — after breakfast I had some more whisky and milk — lovely — we know how to appreciate such things having long felt the want and desire for them — tinned milk was all we had had for two months.

Passed Spencer's Brook where you change for Albany.

About 8 we left for Perth, not the land of our birth, but in our eyes the Canaan we were looking forward to, the land we knew would, metaphorically speaking, be flowing with fish, vegetables, milk, oysters and baths.

We arrived at eleven o'clock, drove to the Grand Hotel and having secured rooms, walked direct to the Barber's Shop, where the baths are, and boiled ourselves. Nothing short of boiling would remove the red dust that had accumulated on us — it is wonderful, it gets through your trousers, socks etc., in fact you are red all over. Then the shower — what a luxury, you must remember I had not had a wash in a bath for more than two months. At the Hotels at Coolgardie, such a thing is not thought of. You can understand when water is 6d. a gallon and sometimes more, baths are not used much. There was one bath in Coolgardie, but it was destroyed in the fire. I fancy Bobby had one, and I think if cost him half a crown, probably he was then using someone else's water with a little fresh added to it.

From the bath we went to the Oyster shop. I can't possibly describe, words fail to express, the supreme enjoyment we experienced in eating those bivalves, which came from Brisbane — first class oysters with a fine flavour, perhaps not as good as "natives", but to our palate they were delicious. Then we returned to the Hotel, put on other clothes and had dinner. They dine in the middle of the day here, tea at 6.30 the same as dinner, but tea in it. We slept well that night; we had a small room each, very small, but comfortable.

November 10.

Sunday morning we took a walk by the Perth Waters through which the Swan River runs, then we

Mount Eliza and the Swan River, Perth.

walked into the Public Gardens, beautiful flowers, and arrived at the Cathedral about 10.45. We had heard that the elder Orchard was leader of the choir. Just before the service commenced we met Mrs Milward, who was on the same deck as Bobby. Her husband was a Mining Engineer at Coolgardie, and we knew him well. In the afternoon we took a row on the Waters, so delightful — you know I always loved the water, and not having seen any for so long, I was quite pleased, and enjoyed it greatly. The two Orchards were with us, and assisted in rowing the boat. I took some bottled beer etc. to refresh ourselves with & we had a fine time. We were much interested in seeing several Pelicans.

The Osborne Hotel.

November 11. Interview with Mr Salmon, manager Bank of New South Wales, a very pleasant man.

On Monday we went to Fremantle on business. On the way back we got out at Claremont, a station on the line, and walked to the Osborne Hotel, which is beautifully situated on high ground overlooking Freshwater Bay, a magnificent stretch of water, and forms part of the Swan River. The bay is surrounded by hills covered with fine trees which grow down to the water's edge. This view is seen from a Tower which forms part of the hotel; it is rather a picturesque structure, after the modern style, having a wide balcony on two sides, overlooking the gardens. There is a very pretty little island* in the bay, well timbered, on which a house has been erected. Bobby and the younger Orchard, who was with us, determined to go down a steep incline to the water's edge. So I agreed to accompany them; it was rather difficult, we had to hold on to rocks and

[* Perhaps Keanes Point, as there is no island in Freshwater Bay.]

stumps of trees etc., but we landed safely at the bottom and scrambled along the rough shore until we arrived at a landing stage, a sort of jetty, where we found some small boys catching crabs, funny round things with blue legs. I don't think they had any claws. They had caught about fifty. I am told they are good eating.

Summer yachting on the Swan River.

Tuesday we drove to "Osborne", a beautiful drive, the first part of it being by the side of the Perth Waters, the remainder by a road with woods on each side. This is the high road to Fremantle, which is about 12 miles, the Hotel being about midway. In the evening we went to a musical entertainment at Mrs Milward's. The two Orchards were there, and a Mrs Hooper, the wife of a mining expert I had met in Coolgardie, a very interesting lady who had lived with her husband for two or three years at "Salt Lake City". She told us a good deal about the Mormons and their doings. We passed a very pleasant evening, plenty of cakes and wine, spirits and cigarettes. Last evening we dined at Mr Horgan's, he is a most hospitable man, and I like him very much.

Saw Shierlow at the Shamrock Hotel, not I should say to be trusted.

November 13 1895. Weighed 13 stone 1 1/2 lb 1903 Oct. 13 stone.

Tomorrow I start for Cue at 6.30 p.m. I am much disappointed at not being able to arrive home in time for Christmas, but what can I do, they have asked me

Started on the18th October.

to go there, and I can't very well refuse I suppose, being in the Country.

I shall return to Coolgardie for the late manager's trial on December 11th, and at the same time carry through the purchase of the new mine. Of course, I shall be paid an increased fee, for remaining longer than the time I had arranged to, and it will have to be on a higher scale as my business in London is being disarranged through my absence.

November 16. On the Swan Waters from 11 to 5 with Bobby rowing and sailing – most enjoyable, beautiful weather.

Bobby has now gone to post a little box containing presents for all the family, including Ernest, who I am so pleased to hear takes care of you and Lulu. We are both well, and should be strong for we have oysters every day. Yesterday I had 3 dozen before dinner, and the same for supper. Bobby had the same number for supper.

I have settled not to take Bobby with me to Cue, he will stay a week at the Cleopatra Hotel, Fremantle, which is the port of Perth. Being by the sea it must do him good, he is sure to have a fine time. Whitworth, our Engineer, is staying at the same hotel, then he returns here to stay with the Orchards.

It is too expensive to take him with me, the coach journey alone being 16 pounds. Two days coaching, I shall have had enough of it by the time I return. Wish everyone for us a merry Christmas and happy new year, and with love to yourself and Lulu.

Your affectionate husband,
Robert Emeric Tyler.

P.S.

About the end of December look out for the S.S. *Rome*, as a case of specimens is on board. I suppose the P & O Company will advise you of its arrival. It will have to be fetched from the steam ship office. Bobby does not wish it open. Thank Louisa for her kind letter, tell her I found a card at my Hotel the other evening, it was Charley Carter's. He called next evening and we had a long chat, he is looking remarkably well, is now in the Government Survey Department. For 4¾. Years he was in the permanent "Victorian Artillery" — Spen & Percy will be interested to hear this — He desired to be kindly remembered to everyone — R.E.T.

Coolgardie
Dec. 11th, 1895.

My dear Emma,

You must be wondering why I have not written you since Nov. 17th, but when you read the following you will understand that I have not had the opportunity. I have passed through much; have travelled very many miles by rail, coach and buggy. But I am still well, strong and hearty.

I don't quite recollect how, when and where I left off in my last letter, the note in my diary is — Sunday 17th November wrote home.

On that day we went to the Cathedral in the morning, and in the afternoon, on the River, we saw three pelicans, most melancholy birds, looking for fish. Until then I was unaware that they patronized Western Australia.

I had settled to leave the following day for my journey to Cue, and I had arranged for Bobby to go to Fremantle — you know all about his stay there, he must have had a real good time.

On the Monday, after providing myself with an umbrella, white outside and green in, a Khaki suit, and other necessary articles, I started by the 6.30 train for Mullewa — 348 miles. 9.15 arrived at Gingin — 12.15 Watheroo, 4.45 a.m. Mingenew — 6.30 Dongara, a sea port — pretty country — very much like English. Saw a flock of wild turkeys — fine birds

— 8.30 Mullewa junction — no refreshment room — 9.30 train stopped for no apparent reason, getting hungry, country flat and uninteresting. 10, another stoppage — why, I know not — still very flat, no trees, only one here and there; some shrubs with very pretty red flowers growing below the leaves. 10.20 stopped at a wayside shanty, had tea and cake — thank God for small mercies.

Mullewa.

About 12 o'clock I arrived at Mullewa, a wretched place, had a wretched dinner at a wretched, so called, Hotel, booked my seat on the coach for a ride of 254 miles, in 14 relays — 1st stage 19 miles, 2nd stage 19 miles. We started about 3 o'clock and arrived at the second stage at 8.15 p.m. Here we stopped 40 minutes for supper, the name of the place was Bunbenoo. (I should have mentioned when the coach started from Mullewa, the 5 horses took fright at a team of donkeys and bolted. Luckily the driver managed to pull them up, otherwise it must have been serious.) 3rd & 4th stages 21 miles. We had then done 80 miles, and had come to a place called "Chainpump".

Gascard coach proprietor.

Return fare £15.0.0. Luggage 10/- Driver 10/-.

November 20.

We arrived there at 4.30 a.m. and left at 6.30 after having a good breakfast. 5th stage 20 miles, landed us at Yalgoo at 9.30 — here I met a man named Faddy who had come down with us in the coach from Coolgardie. Left at 10.40, 6th stage 20 miles — very pretty country, like riding through a park, started on the 7th stage 15 miles at 1.45 — arrived at a well at 3.50, very good water — left at 4.20. 8th stage 25 miles, arrived at the stopping place for the night at

7.45, very rough canvas tent, bad food, unfit to eat, slept in the tent, three bunks side by side, very primitive, turned out at 3.30, the first up. This was Thursday — left at 5.15 9th stage 15 miles, 10th stage 20 miles — Mount Magnet 10 a.m., excellent lunch. Hotel kept by a man named Attwood. 11th stage 16 miles, 12th stage 12 miles, 13th stage 28 miles. Eight miles brought us to Lake Austin — a waterless lake — stopped at the Island Hotel.

November 21.

Arrived at Cue at 7.30 after passing through "Day Dawn", a town called by that name, 3 miles from Cue.

Cue, population 1902 about 600.

I think there could only have been 13 stages, because the distances given make up 254 miles.

It was an awfully fatiguing journey. I had the box seat, but sometimes I was so tired, that I was obliged to get an inside passenger to take my place. At times I felt that I should fall off the coach onto the horses, a very unpleasant sensation. My fellow passengers were not high class — in fact, they belonged to the working class, except one who was a traveller in some line of business, his name was "Thieru", nephew of the Thierus of Glasgow and Edinbugh — this was rather strange. There was also one girl who was on her way up to the hotel at Day Dawn.

Mr Carpenter met me on my arrival. On gaining the hotel, I found it full, a dinner was being given to the Minister of Mines (Mr Wittenoom). I changed my clothes, that is, put on a clean tennis shirt, and clean pants, and attended the dinner. I thought I might as well hear what was going on in the mining centre of the Murchison District.

Marks, hotel proprietor.

Cue. Capital of the Murchison Gold Mining District.

It was a mutual admiration society — everyone patting everyone on the back, and one and all abusing Coolgardie. I was glad when it was over. I was, as you can imagine, about tired out. I don't know if it was a good dinner, for the simple reason that I could get nothing — about 4 waiters to 50 guests, and they were nearly drunk — mostly all the guests (including the chairman, who was the mayor) were in their shirt sleeves, but I think Mr Wittenoom, being a member of Parliament, was in a black coat.

Introduced to "Tom Cue" the discoverer of the Cue gold district. November 22. 1900 he was leaving Victoria Station for South Africa.

Mr Gale.

I sat next but one to Mr Wittenoom and had some conversation with him on mining matters: he is now (1898) the representative of the Colony in this country.

After some difficulty the land lady — a very nice sort of woman — found me a bed in a small room off the back yard. There was another bed in it, which was occupied by a man called "Camel Thompson". I never saw him awake — I was in bed and asleep

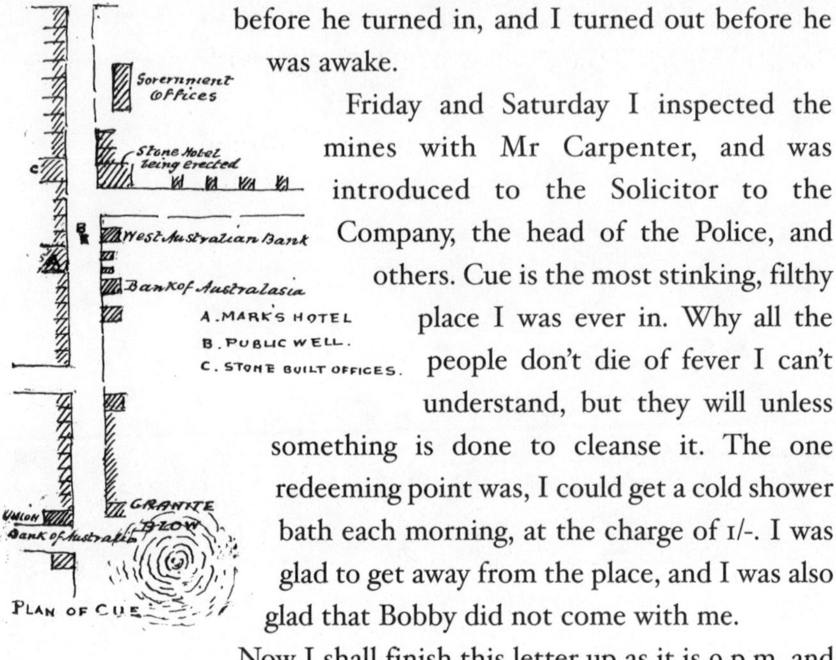

before he turned in, and I turned out before he was awake.

Friday and Saturday I inspected the mines with Mr Carpenter, and was introduced to the Solicitor to the Company, the head of the Police, and others. Cue is the most stinking, filthy place I was ever in. Why all the people don't die of fever I can't understand, but they will unless something is done to cleanse it. The one redeeming point was, I could get a cold shower bath each morning, at the charge of 1/-. I was glad to get away from the place, and I was also glad that Bobby did not come with me.

Now I shall finish this letter up as it is 9 p.m. and the mail leaves at 10, so I shall reserve my journey to Pinyalling for my next letter which I hope to write at the end of the week.

As you see I am now back again in Coolgardie, I arrived yesterday morning. Today Lowden's trial came off, the jury disagreed, and I declined to go on with the prosecution. So, thief that he is, he is a free man again.

Tomorrow morning at 4.30 I am leaving for Hannans. Bobby is going with me. I should think now in a very few days I shall be making tracks for home. I want to get there to see you and Lulu, and all my relations and friends. They cannot realize what I have passed through during my visit to Cue and back. I did 700 miles of coach and buggy driving, and 618 miles of railing. Besides the journey from Perth to this place — 310 by rail, and 41 miles by buggy. With love to all.

Your affectionate husband,
Robert Emeric Tyler.

Austin Street, Cue, the main thoroughfare.

Coolgardie — 8 p.m. 94° in the office —
112° in the shade — 180° in the sun
Dec. 14th, 1895

My dear Emma,

I think my last letter finished up with my arrival at Cue, and the dinner I attended in honor of the Minister of Mines, Mr Wittenoom, that was on Thursday November 21st. On the following morning I went with Mr Carpenter to see the Mines belonging to the Company. The Murchison Gift was within 10 minutes walk of the town. Apparently it is no good. I fancy it was formerly called the Christmas Box, and the prospector of it, that is, the man who found it, either fell down or threw himself down the shaft, anyhow he died. The second property, "The Polar Star" (rather a cold name), was about two miles distant, or probably not so much — appears to be a really good mining property, and the reef shows good payable gold. I think it is a good property, and I am bringing some specimens from the mine home. I'm a collector.

The following day, Saturday, I again visited the mine. Whilst I was there a tiger snake & very dangerous, was killed by the underground Manager (Cairns) in the room next to the office, where Mr Carpenter will sleep when the place is straight. The galvanized building, consisting of two rooms, had only just been completed, in fact two men were covering the roof with stuff cut from the bush whilst I was there, to form what is called a "fly" — it keeps the great heat from off the roof. The snake was over 4 feet long. Just outside the building we found the hole it came out of, and as they usually run — or rather wriggle — in couples, we put a dynamite cartridge down the hole. The effect when it exploded was fine, the dust went up like a column into the air. What became of the snake inside, history does not relate; before I left one of the men had skinned the snake, I watched the interesting occupation.

The horses came to grief outside the gate leading to the Golconda Mine — it was the only mine I saw fenced in.

We then returned to Cue and, having dined, left at 6 p.m. in a buggy and pair, for Pinyalling. During the morning we had purchased all the stores we required for our journey, and secured the buggy and pair which was to take us there.

I was quite glad to turn my back on Cue; I never could imagine it possible that a place could stink (that is the only word to use) so badly — it was too horrible

Cue Volunteer Fire Brigade Station.

Lake Austin Gold Mine Battery on Lake Austin.

— the place had only one redeeming feature, you could obtain for 1/- a cold shower bath, which was a luxury indeed, this I mentioned before.

The hotel was situated on the "Eureka Mine".

In 1855 Richard Austin predicted that the Murchison around the shores of the salt lake that bears his name would become one of the richest gold fields in the world.

At 9 p.m. we arrived at the Island Hotel, Lake Austin, 20 miles, had some supper, and turned in.

Sunday November 24th, a most eventful day — I was up at 5, and by 7.30 the horses were ready to start. We had partaken of a good breakfast, and loaded up the buggy; I was standing between the wheels, putting something into the front, when I felt the buggy beginning to move. I stepped back quickly, the horses broke into a trot, then into a gallop. Carpenter tried to stop them, but it was no good — they bolted over a large quantity of bottles, covering a space of about 30ft square, a heap that had been accumulating ever since the hotel was opened perhaps 5 or 6 years ago, then down a steep hill and were lost sight of.

Carpenter and a man went after them; they were found all of a heap at the bottom of the hill. The next I saw were the horses being led back, poor brutes; they were in a shocking state, dreadfully cut about the legs and presenting a most miserable appearance. Now this unfortunate accident could have been avoided — the waiter, or general man at the hotel, was standing practically in front of the horses — he never made a movement, or so much as raised a finger to stop them. I was so annoyed that I abused him, in not quite parliamentary language, and the Landlady of the Hotel sent him off then and there, without even his breakfast.

The horses came to grief outside the gate leading to the Golconda Mine — it was the only mine I saw fenced in.

Carpenter, although not understanding much about horses, purchased yards of linen which he saturated in some carbolic stuff. This he wound round their legs, sewing it on in a very masterly fashion, but the grey — one was grey, the other brown — lay down and everyone — the whole town was out, it being Sunday morning — thought he was going to die, but after a few hours he managed to get up and take something to eat and drink, and began to mend. The brown appeared not to be quite so bad. The horses were in a sort of straw yard — or rather a large pound, enclosed with high posts & rails — there being no stables, all the horses are turned into this enclosed space.

You may imagine I was in a fix   20 miles from any place where fresh horses could be obtained. In our difficulties we appealed to our Landlady. She advised us to see Capt. Williams of the Austin Mine.

Mrs Philburn.

Capt. Williams was leaving the Austin mine to manage the "New Chums" at Magnet where he remained until February 1899. The shaft on this mine is 1000ft. deep.

He is a good man — she said — and will help you if he can.

His place was about ½ a mile from the Hotel, just beyond where the horses came to grief. It was not long before we were there. I knocked at the door, and out came Capt. Williams. I said — Well Captain, we've met with an accident, our horses bolted and have smashed up themselves, and the buggy, and we've come to you to assist us. What do you want? he said. Well first we want a Blacksmith to mend the buggy. You can have him he at once he replied. What next? We want the loan of your horse, and a man to ride into Cue with a letter to the Livery Stable Keeper to send us fresh horses. I can arrange that, he said, but you'll have to pay the man for his time. Of course, we were agreeable to that. He then found the blacksmith who worked until dark on the buggy, and made it as safe as before the accident. But after all we did without the horse and man, as a young man, "Sammy Marsh", the driver of a coach, and who was driving into Cue, took the message for us. In the evening two fresh horses arrived, at which we were delighted.

Camp of the Inspector of Mines, Cue.

I think I must call this "The adventures of two men, and a dog, in a buggy" for we had a little white English terrier, belonging to Mr Carpenter, with us. She was a nice affectionate little dog. All through the journey she was on the seat between us. At the hotel I met a Mr Dickinson, manager of the New Chums Mine at Mount Magnet, and in the course of conversation I found he went on board the Ormuz at Albany as I came off, and took possession of the very bunk I had quitted. He of course knew all about the people on board, who I had been with for four weeks, so naturally it was interesting to me to talk about them.

Monday, November 25th. At 7.45 we started, this time we got safely off, with a much quieter pair of horses, no bolting about them. At 11.10 we stopped for lunch, tinned Oxford sausages, biscuits, and whisky and water. I again met the nephew of the Thierus, he was on his road up by coach to Cue; whilst the horses were being changed he came and refreshed himself with us.

Lunched at the "Caves". There were some caves at the side of the track and water hole close by.

Starting once more we made for "Badges Occidental Mine", and interviewed "Badges", an aged man with spectacles. He presented me with a specimen from the reef out of which he hopes to make his fortune — then we called at the "Rock of Ages" mine. The manager Mr Blackburn treated us to whisky and water — a good sort of man much to be appreciated in the bush. At 5 p.m. we arrived at Warramboo — Mount Magnet — 44 miles from Cue, and put up at Attwoods Hotel, not a bad one for that part of the world. We then walked to the "New Chums" and had dinner with Mr Dickenson, and

Had an unpleasant adventure with the Landlord Attwood through his wife, a drunken woman.

New Chums – 400 tons produced 1100 oz.

looked over the mine, which is considered a good one. Before going to bed, the landlord was good enough to let us have a bath, which you may guess was a luxury.

Tuesday November 26th. We were unable to get away before 9 o'clock, owing to the difficulty in getting fodder for the horses. We were about to be leaving the main track, so were obliged to provide ourselves with it. Eventually we obtained 112lbs of compressed fodder for £2.11.4, also some oats and bran; when this was fixed up at the back of the buggy we were well loaded up. Before leaving we had really a good breakfast of chops, kidneys and bacon.

Yowaragabbie 58 miles from Cue.

At 12 o'clock we had arrived at Yowaragabbie (18 miles), this word means, in the native language, "Red Water". We took a spell here of 3½ hours, having our midday meal, which we very much enjoyed. Mr Blackburn drove up just as we were about commencing. He luckily had some chops, which he cooked over a fire we made, on a bit of bent iron. This answered the purpose of a grid iron; the fire was made of a few sticks on the ground, the iron being placed on the top, but although the meat was burnt black, it was sweet, and good. Here I met the brother of Warden Finnerty of Coolgardie. I saw a tall man coming out of a cottage — the place we had stopped at was a sheep station — at first I thought it was the warden himself, the likeness was so extraordinary — I said, Is your name Finnerty? To which he replied — yes. Well, I said, you must be the brother of the Warden at Coolgardie — he was. We opened a bottle

Mr Blackburn called at Clifton House Osterley, July 25 1907.

Watson's Station.

of whisky, and he stayed with us until we left; he was a very jolly fellow and we passed a pleasant time. He was a Government official, and his business was to drive about the country looking after "scabby" sheep.

Mr Finnerty said tell my brother "I'm going to make a spoon, or spoil a horn"

It was fortunate we met him, as he gave us a letter of introduction to Mr Barney Lamond, the manager of "Baron Rothschild" at Pinyalling, the place we were making for.

Having said goodbye to Mr Blackburn and Mr Finnerty, we started off at 3.30. Half a mile from where we had camped, we had to leave the main track, our instructions being to follow some buggy tracks going in a southerly direction into the bush. At about 18 miles we should have come to an old shed and a well, where we were to have camped for the night, but unfortunately we missed them, it came on dark. I walked in front of the horses for about six miles, thinking we might strike the well, but when it got to 8.45, we gave it up, took the horses out — poor beasts they had to do without water — fed them well, and having had some Bovril on biscuits and "billy" we commenced preparing our camp. To cover the things at the back of the buggy we had a large tent cloth — this we fixed on to the top of the wheels of the buggy, then spread it on the ground; on this was placed a waterproof rug, and I utilized the seat of the buggy, so it was not quite so hard as lying on the absolute ground. My portmanteau I arranged as a protection for my head.

Fogarty's Well.

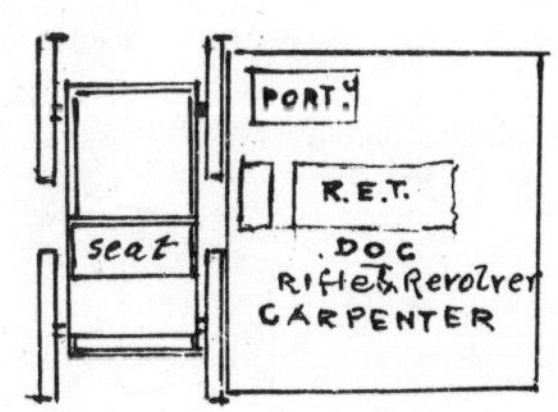

I provide a plan which will assist you to see the position. I then rolled myself in my rug, a long one which I had purchased in Perth. I should mention that we had a rifle, and a revolver; these we placed between us — nothing like being prepared. Fortunately we had no necessity to use them.

We lay down at 9.45 — the little dog sleeping between us. In the middle of the night I woke; it was cold, and so quiet, not a sound of any description, it was a new sensation, to be sleeping without any cover, but the starlit sky. A feeling came over me difficult to describe, it was my first experience of real camping in the bush. As I lay gazing up at the stars, I thought of all those dear to me so many miles away, and thought and thought and thought, until I slept again.

At 4.30 I got up, and by 5.45 we had fed the horses and were on the road again (this was Wednesday November 27th), hoping we should soon come to a well, so that the horses might have water, which they evidently wanted, not having had any since the previous afternoon.

The first well we came to had no bucket, or rope, so we had to push on, much disappointed. However, at 7.30 we came to a well, which had a rope and bucket, and soon the horses were satisfied, and we had a good wash. We had not been short of water, having our water bag. The sketch shows the well

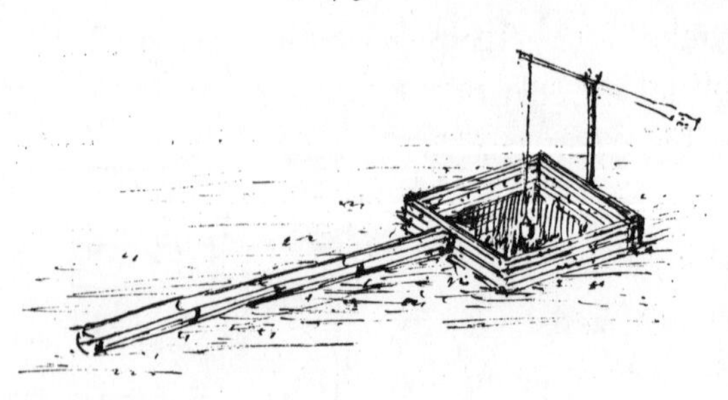

and the trough, a tree hollowed out from which the sheep drink. The principle used for raising the water from the well, is the same as was used at the time of the early Egyptians.

I was very much interested the previous day in seeing some Kangaroo; the way they jumped and the rate they went at was extraordinary, the biggest must have stood over 6 feet high. We also passed some Emu. At 9.10 we met two carters who directed us to the Jarraminda well, which we arrived at about 9.20. At 9.45 we crossed a waterless lake — we were then about 43 miles from Yowaragabbie. At 11 a.m., we came to another well, took the horses out of the buggy, had another good wash and a shave — fancy shaving in the bush — we then had hot Bovril, tinned mutton and whisky and lime juice. After that we went to sleep under a mulga bush — the name of this well was "Mardagullamarra" — you must have a good memory to recollect these names.

At 2.30 we harnessed the horses and started for "Clinch's Station", a sheep farm, where we arrived at 4 o'clock. After taking the horses out, and feeding them, I purchased a loaf of bread from the wife of the owner of the station, for which I paid one shilling — quite enough too — and witnessed for the first time sheep shearing — I pitied the poor sheep. I also purchased some mutton, but had to wait until a sheep was killed. We however supped with the Squatter and his entire family, including children and labourers on the station, all seated at one long wooden table — no cloth — and seated on forms.

The supper consisted of liver and other parts of the interior of the sheep — it was beautiful! The station consisted of 97,000 acres.

Imagine yourself on a plateau, which appears to be situated on the top of a globe. You look around, and see a perfect circle, the horizon alone broken here and there by the scrub, and towards the west by slightly rising ground.

The house in which the squatter lives stands in the centre of the circle, on a red sandy plain, facing the west, and consisting of two rooms, and a long room attached, in which the shepherds have their shakedowns. It is a one storied structure with a verandah in front, up the supports of which sweet peas are endeavouring to climb, and the openings in which window sashes should have been, are covered with canvas. It is built of blocks composed of burnt clay and sand, the roof being covered with galvanized iron — 'Tis not a lovely picture, but true to nature.

All had the appearance of decay and ruin; three years of drought had brought its curse. At the back of the house stood the well for drinking purposes, and to the north the well and trough for the sheep and horses. Early next morning I saw a great number of grey parrots with pink breasts drinking at the trough. We camped by the well at the back of the house. This place was called Bunnerbinnar (red ground).

Thursday, November 28th. A fine bright morning, we turned out very early (that is hardly correct, as we had nothing to turn out of), and by 7.35 we were ready to start. The squatter, Clinch, volunteered to put us on a near road to Pinyalling across Lake Monger, the same route a buggy had gone a few days before. Having started us fair on the buggy tracks, he left us to our own devices. It took three hours to cross the lake — of course, a waterless one — it consisted of a bit of lake and then a bit of island, and this went on for the three hours, we thought we were never coming to the end of it. Once we missed the track, Carpenter had to get down, and walk backwards and forwards until he discovered it — it being sandy ground, the wind and the sheep had obliterated the marks. This kind of travelling was most fatiguing for the horses, besides which the heat was intense, over 100° in the shade.

At 2.15 we were nearly at the foot of Mount Pinyalling, which is a range of hills running nearly north and south, the southern point finishing abruptly at the edge of a part of Lake Monger. We had to find a "soak", that is, a hole in which water collects that comes off the hills. We were unable to discover it, and after driving in all directions for an hour, we took the horses out to give them a rest; it was awfully hot, and they wanted water badly. We started again at 4.30; for a long time we were unable to find it. At last I said, we will keep on the track until we come to the point where the hill stops at the lake. The soak was there, and glad we were, the idea of being bushed was far from pleasant & I must admit I was beginning to get somewhat excited. It was then 5.30.

Pinyalling Soak

The soak was at the very foot of the hill, within 60ft of the lake. It was simply a hole about 6 feet deep with stones piled round, about which some scrub had grown. I saw it first, and sung out to Carpenter, who was looking about trying to find it. We lost no time in giving the horses a long drink; it was fresh water, in fact the rain water that fell in the hills had percolated through the ground. It was alright for making tea, but not very good for drink purposes.

When about commencing our supper, a little Aborigine boy rode up with five horses to water at the soak. He was riding one, and had a long whip with a short handle, a regular stock whip, with which he kept the remainder in front of him. We immediately made friends; his name was Willie, and he was only ten years old, quite an interesting little fellow.

At that moment a terrific storm came on, without the slightest warning. Any amount of thunder and lightning but not much rain, it was grand whilst it lasted; we all three dived under the buggy for protection. We then entered into an arrangement with Willie — he was to stay with us until the morning, when he was to bring in our horses, which although hobbled, would probably have strayed away some distance. He was to have a shilling, and his "tucker", and slept under the buggy on a sack with another to cover him. We cooked over a fire we had kindled the mutton chops purchased at the last station, and made our "billy" which we heartily enjoyed. Then, preparing our camp as usual, notwithstanding the ground being covered with ants — not small ones as in England — I slept right away from 10 to 4 a.m.

The sketch will give you an idea of the spot where we were camped. Immediately in front of the buggy was some low scrub, then the waterless lake, with scrub & trees on the other side. Beyond came a low range of hills, the highest being called the Sisters, and in the far distance, some thirty miles away, Mount Warriedar, which had a soft grey appearance. It was a fine view — the dotted lines indicate a track across the lake leading to a mine called "Field's Find".

It was from this mine that Willie had brought the horses. The width of the lake was about a mile. Camping not far from us, was a man with four pack horses. He lent us a frying pan, which came in useful for the chops; all the people you meet in the bush are most obliging. They see so few people, that when they come across anyone, they like to have a talk about things in general. Whilst we were discussing the chops, he threw himself down by us, and talked, and we gained some information as to the way to get to the Baron Rothschild Mine; he sketched in my book the route we were to take.

Friday November 29th. Up at 4.10 — it was a fine morning after the storm, made the "billy" and cooked more chops whilst Willie hunted up the horses, and brought them in; he then went after his own, and whilst he was away his Master rode up, we saw him coming across the lake, he was just wrathful, and abused us for keeping the boy all night. However, after a bit he calmed down, and I don't think Willie "tasted" the whip — at least I hope not. We gave him the promised shilling, but after returning with the horses, he said he had lost it — we rather doubted the statement — he was a "cute" boy and very curious, as the following shows: whilst standing in front of me, he commenced prodding me in the waistcoat, or where it would have been, had I had it on, with his finger. I imagine he wanted to find out if it was all real.

Field of "Field's Find".

At 9.45 we left the soak. We had the plan the man had made, and we felt happy and confident that we should reach our destination in about two hours. But fate was against us, we drove for an hour in the direction shown by the plan, and then came to the conclusion that it must be wrong; there was nothing to be done but to turn back, which we did, arriving at the soak just as the man with his pack horses was moving off. We were in luck this time. Had we missed him, I don't know what we should have done — the plan, he at once admitted, was wrong; instead of going to the left, we should have gone in the opposite direction. He then put us on the right track, and we were thankful. The road we had to travel passed straight over the Pinyalling hills, the mines being on the opposite side.

At 2.15 we met a man riding, who was going to the Rothschild mine to get a sheep, so we pulled up on the top of one of the hills and had lunch, the man joining us in a pannikin of tea, he then piloted us to the Mine.

The view from this spot was simply grand. We were surrounded by deep valleys, and hills, as far as the eye could reach. The whole covered with trees, in fact it was one dense forest of fine trees and scrub, but no big trees.

It was 5.15 when we arrived at the Baron Rothschild Mine. I at once handed my letter of introduction from Mr Finnerty to Mr Barney Lamond, who received us in the most princely manner. He rose from the couch on which he had been reposing, after the labours of the day, stretched himself to his full height — he was over six feet — and held out the hand of welcome. He was dressed in rather a dirty suit of Pyjamas, torn in various parts; he was not strictly speaking handsome, but had a pleasant smile, and happy face.

He had but nine fingers, having lost one in a fight with some Afghan camel drivers. Whilst driving a team of fourteen horses, the Afghans refused to move their camels off the track, and when remonstrated with, they attacked him with sticks.

He landed one in the mouth, dislodging a few of his teeth, but unfortunately his finger was cut, blood poisoning set in, and his finger had to come off. He had fought the niggers, or Aborigines in Kimberley — northern Australia — and had been shipwrecked on the coast of Australia, in fact, his had been an eventful life, and after the dinner I am about to describe, he told us many interesting and amusing stories. But the dinner! Directly he had welcomed us, he commenced gathering sticks, lit the fire, and put the billy on, and a saucepan. I wondered what could be in the saucepan; it turned out to be sheeps head soup — delicious — the chops were excellent, likewise the tea. Everything was good, then he handed us a fine cigar, and as I have just mentioned, told us many exciting anecdotes.

At 9.45, we again camped by the side of the buggy — a bed he was unable to offer us, his tent only containing one, and that made of bush timber. I should have mentioned, as we were sitting down to dinner, four Indian girls came in from tending the sheep — one not bad looking for an Indian (it is a very ugly race). They regaled themselves on our leavings, after making a fire not far off, on which they heated their "billy" and warmed themselves, which I thought they required, seeing that their only clothing consisted of a short — a very short — sort of night gown. At night, I believe, they have a blanket — the rather good looking one was a swell, she wore an old yellow bodice. If it is not too indelicate, I might draw attention to their legs — well, I never saw such legs

Team of seventeen donkeys drawing five tons of water.

before, absolutely no shape about them, in fact no more than bits of stick, about 4 inches in diameter.

Saturday November 30th. At 4 when I awoke, the stars were shining brightly over my head as I lay on the ground. At 4.30, from my recumbent position, I could see the sun beginning to show above the horizon. The effect was strange, to see the stars, and the sun at the same time. Whilst stripped to my waist washing, two Indian women (old ones) came up to the camp, each with a bit of wood smouldering. They soon made a fire, and sat warming their bare knees. Poor things, they must lead a wretched existence, but I doubt if they have the brain capacity to appreciate their condition.

By 6.30 we had finished our breakfast, then Barney Lamond showed us round the Mines, which were all clustered together. The Lady Mary, the Bute, the Rob Roy, the Baron Rothschild, the Portland and

Col. North was interested in the Baron Rothschild. I had an interview with him about it an hour before his very sudden death.

the Dublin. This was the mine Mr Finnerty was interested in, he had asked me particularly to look at it, which I did, and I saw there, that which is seldom seen — a "Jew" working with a pick. Whilst standing on the brace of the Rothschild mine, which was half way up the side of a high hill — and from which I had a grand view over the valley and country round, two dark objects were pointed out to me. I asked what they were, and was informed that they were our horses that had strayed away. A black fellow was sent after them, and soon brought them back. These black fellows can track anything in the bush, from man to a horse or camel — it is wonderful they can detect on the ground a mark where we could see no sign of one.

We returned to the camp at 12, had a good lunch or early dinner — more chops etc. etc., and started at 1.30 for Yalgoo where I had to catch the coach. Barney Lamond is a real good fellow and treated us in the most kind and hospitable manner. It is said that he is one of the greatest liars in W.A., or for the matter of that in the whole of Australia. But then, most men are liars in this country, more especially those interested in mining operations.

By 5.35 we had done 19 miles, the horses went well, I suppose because they knew we were making for a "well". We had passed some clay pans, and gum trees, the first we had seen in that district; it was a most lovely afternoon, cool and beautiful. The roads about this particular part were not simple tracks, but good hard roads, about 10 feet wide, made of red gravel.

Well, they looked as if they had been made, but I don't imagine they had been, of course. This being in the district of sheep stations, it was considered necessary to have fair roads for communication with the sea port — this is my idea, and probably the right one.

About Pinyalling and the district we had driven through, there was no big timber — fair sized trees with the leaves growing at the top, which threw but little shade, and low bushes. When we were searching for the "soak" and took the horses out for a rest we had the greatest difficulty in finding a shady spot to put them under. We crawled beneath a bush which protected us a little from the great heat of the sun. "In this country the trees cast no shade — or very little — since every leaf is set at edge against the sun, and shed, not their leaves, but their bark, which, stripping off in long scales, exposes the naked wood beneath, and adds to the ghastly effect which the forest already holds in the pallid hues of its foliage."

Meleyer Well.

Whilst at the well, in the distance, we saw a man skinning a sheep, and some others camped close by. On driving up we found that one of them was the owner of the next station — Mr Eyken. Providence administers to the shorn lamb, and also to the hungry man. We purchased a leg of mutton for 1/6 — not dear you'll say, and with that in our larder, or rather I should say our "boot", we drove on to Eyken's station where we arrived about 6.10.

The house was closed, all but the kitchen, and one room left open for any weary traveller passing by to

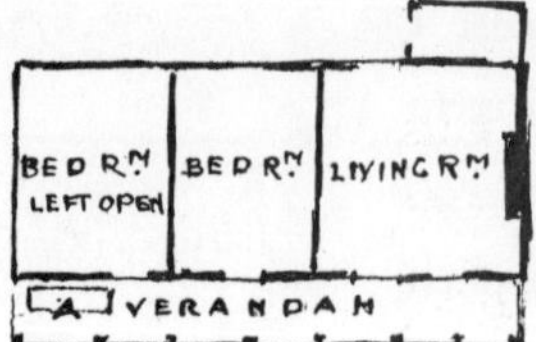

sleep in — at the most there were only two more rooms, all on the ground floor, but it was far superior to Clinch's station.

Another traveller had just arrived, Mr Williams — he was a government sheep inspector, on the look out for "scabby" sheep — not a pleasant occupation, but necessary, one sheep suffering from it will communicate it to an entire flock. I found a frying-pan in the kitchen (thoughtful man, Mr Eyken) and cooked some of the mutton for supper — which we enjoyed. Instead of camping at the side of the buggy, I slept on a shakedown under the verandah of the house.

Sunday December 1st. I woke at 4 o'clock, and turned out at 4.30 — it was a lovely morning. I went to the well, which was situated some distance from the house, and had a real bath in the sheep trough, so pleasant and refreshing. I should mention that we were now in a district where water was to be seen, small lakes and ponds, you can't imagine how delightful it is to look on water, after you have been travelling through the bush and in a waterless country. For breakfast I cut from the centre of the leg of mutton three steaks, which I cooked with some onions, carefully sliced, which I had discovered in the kitchen. Well! It was just fine. The sheep inspector joined us. From him we obtained a plan of the route we had to take, and at 8 o'clock we were again on the road.

At 8.30 we came to the first sign of real grass. In the other parts, the sheep feed on the salt bush, and a sort of weed that grows in the sandy soil. From what I saw, they don't seem to get very fat on this fare, but they are only bred for the wool — at Clinch's we saw them doing the wool up in great bales ready for despatching to the nearest mart, which I suppose would be "Geraldton", from whence it is shipped, most probably to England.

On our journey we had passed many deep creeks and water courses, no water in them, but showing undoubted signs of mighty rushing waters, the banks torn about in the most extraordinary manner, and uprooted trees.

Our friend, the sheep inspector, informed me that eight years ago a most terrific storm took place, which inundated the entire country, thousands of sheep being killed, and many lives lost, probably amongst the Indians.

This part of the county was most picturesque, great granite blows, or hills, cropping up in all directions.

At 11.45 we arrived at a rock hole, bad water, but good enough for the horses; here we took the horses out for a spell, and lunched. There were a great number of hawks about, fine big birds — we tried hard to shoot one. Carpenter shot four times, I shot

twice, the bird returning each time to the same twig, on the same tree — it was rather a treeless part where we had camped, but after the sixth shot, he must have imagined it was tempting providence, so returned no more. The barrel of the rifle must have been crooked!!

After a rest I mounted a large granite blow, and on the very top found a small indentation, covering a space of not more than 3 or 4 square yards, by a few inches deep, with water in it. One would have imagined with the great heat it would have dried up. I also discovered the remains of a dead horse. I took a small pick or hammer with me, thinking I might do a little knapping, the result being that I landed a blow on my thumb, which made me dance round for a few moments.

Thoroogycogy Well

We were off again at 2.45, passing through a very stony country, the track being nothing but big stones, very difficult to drive over. The next well was in a hollow. Whilst I was getting the water for the horses, a flock of thirsty sheep arrived. As a good Samaritan I commenced filling the trough, but on looking round, found that one of them was on his back in the trough. I promptly had him out by his four legs; the Indian Shepherdesses — four of them in scanty attire — then appeared, and we gave them the remains of a box of sardines, and some biscuits. They seemed to enjoy this very much. One of them said to Carpenter — "You gold fellah", "You come along Pinyalling", showing that they knew where the gold was. They then put us on the right track for Morris Station —

Muralgarra, where we arrived at 6.15.

The man occupying the station was a manager, and both he and his wife were very civil to us. When we drove up, the wife and an aged man were carrying between them a large iron tub from the well full of water; in her other hand she had a pail, also full of water. I jumped out of the buggy and carried the pail, result — all we wanted, including sugar which we were in need of. I purchased half a cold leg of mutton and some bread, which we thoroughly enjoyed, having tinned oysters first — dangerous but very good — again camped by the side of the buggy.

Monday December 2nd. Woke at 4.30, the moon just dipping behind the distant scrub, and the sun commencing to appear in the East. It was very cold — in fact had been so all night; it woke me up.

All through the journey it had been my business to feed the horses — it was a business. The compressed fodder was more like paving stones, it came out in slabs of about 2 inches, then it had to be broken up, and mixed with the corn and bran — this took time. I also had to attend to the horses; both of them had open sores on the withers, that is, on the shoulders where rubbed by the collar.

The landlady at the Island Hotel, Lake Austin, had told me what to do, which was, to make a lather of common yellow soap — some of which she gave me — and apply it, night and morning, to the injured parts. It was a capital remedy as it kept the flies off — without it the sore parts would be black with them in a minute.

That morning I had a nasty pain in my back and chest; I was unable to stoop down to fasten up my portmanteau — I didn't like it a bit. We left at 5.45, arriving at the first well at 6.25. I should mention that the previous evening, a young man in the station kitchen drew in my book a plan of the road we had to follow to get to Yalgoo. He knew all about it, so he said, but not quite enough, as the result will show.

Mumboojuboo Well or spring.

At this well we saw a flock of green parrots, lovely birds. 7.27 pulled up at a soak and had our breakfast, then drove through a very rocky country. This means not only that the general character of the country was rocky, but that you have to drive over stones and rocks, the jolting from which is no doubt good for the liver. The day was very hot indeed, consequently the sun answered the part of a plaster, and drew all the pain out of my back.

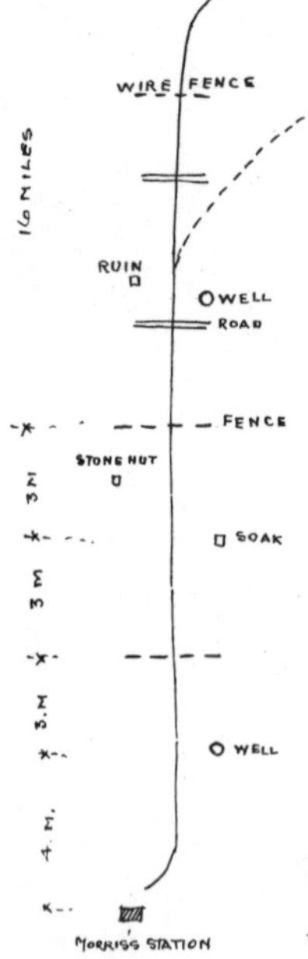

This day's journey was reckoned to be 29 miles, and we expected to be in Yalgoo at 1.30, in time for dinner. But no such luck, the plan was not quite correct, the young man had shown a straight road with a turn to the right about three miles from Yalgoo — instead of which, at a distance of about 21 miles, we should have turned off. As a matter of fact, I did drive some short distance on the right road, but Carpenter said we must be wrong, as the plan showed a straight line — so I turned round and followed the straight track. Well, we drove on for some miles, but not coming to Yalgoo, we took out a plan of the country that I had with me, and the compass, which proved that we were going South West instead of North, then we knew we were wrong, and felt inclined to curse that young man. At last we met a Carter who said that Yalgoo was about ten miles, we were to drive on about seven miles, when we should see a track bearing to the right which we were to follow.

This man was also a fool, or a liar — we drove on and on, until not finding the track to the right, and on referring again to the compass, we determined to turn back and strike a road going north. About 2 we gave the horses a spell, and had something to eat. We didn't relish the position at all; we were fairly bushed, had no idea where we were, or where to find water for the horses; we had some remaining in the water bag. I had named the horses "Billy" and "Possum"; they had both had enough of it, especially Possum. We started again at 3.45, keeping as near north as we could — we were not then following any track. After driving some time in the hopes of coming on a track leading north, we were fortunate to hit upon a clearing which had been made through the bush. It was not more than 6 feet wide, and extended a considerable distance. This we determined to follow, but could only do so at a walking pace, there being but just sufficient space for the buggy, Carpenter walking in front of the horses. At last, to our unspeakable joy, we saw a windsail (a canvas ventilator to admit air into a shaft) in the far distance. It was like a sail to a shipwrecked Mariner. The Mariner would no doubt have said Saved! Saved!, that being usual on such occasions — but if we didn't say it, we thought it.

"Bushed."

The Joker Mine. (Turned out a duffer.)

The mine turned out to be the "Joker", supposed to be a good mine, just sold for a large sum. The people at the mine soon put us on the right track; it was eleven miles into Yalgoo, and such a road, winding by the side of hills, and through valleys, and nothing but stones. "Possum" was so disgusted and annoyed that he nearly gave up the job and would have done, I believe, had not "Billy" kept on biting at his ear. "Billy" was a good horse, but then, "Possum" was fat, consequently not so energetic.

Yalgoo goldfield proclaimed February 8 1895.

It was 6.15 when we drove into Yalgoo, and we had been on the road since 5.45 a.m.; at the very least we had driven 50 miles, if not more. It was a fatiguing day for both man and beast, besides the anxiety as to whether we should be bushed. I had driven all day, so you may imagine I had had quite enough of it, but it was an experience!!

Viscount Avonmore (Yelverton family) held a publican's licence in Yalgoo: the motto of the family – "They will rise again." Let's hope they may!

We dined at the so-called Hotel, off a sort of Irish stew flavoured with curry — well! — I thought it simply delicious. With it I drank two bottles of good English beer, I felt I required something sustaining.

We again met Mr Blackburn, who lunched with us at Yowaragabbie. He knew the brother of Mrs Sutton-Sharpe — the wife of Jon Sharpe's son. The name was Leigh I fancy, funny was it not!

As the Hotel was full, we had to sleep in a hovel, a dirty stinking hovel, but before we could take possession, we had to turn out three drunken men. At first they objected to go, but out they went, right was might, and we wanted to turn in.

Tuesday December 3rd. At breakfast I met

"Sammy Marsh", who I mentioned before. He was driving a new coach between Yalgoo and Mingenew, about 115 miles; when I found that I should pass through two new mining centres, I determined to travel with him instead of going by other coach the following day to Mullewa, besides it saved 150 miles of railway. He was leaving about 3.30.

After breakfast, with Mr Blackburn & Carpenter, I visited the "Emerald Mine", 7000 oz of gold were taken out of the hole, about 30'x 20', by 6–7 ft. deep. I inspected the hole. They have erected a good house for the manager, and put up machinery, but no more gold has been discovered; it was only a patch. The find was made by an Indian woman, who of course made nothing by it.

The manager was a good fellow; he was delighted to see us. He came to W.A. a non-drinker, but he had to give it up and take to whisky, it was no good, he felt he was obliged to have something to enable him to keep up his pecker.

Mr. Porrett.

Mr Blackburn then introduced me to the Warden, P.L. Gibbons Esq., we had a pleasant chat, mostly about his district and mining. I suggested sign posts being placed on the outlying tracks, and related my experience of the previous day. He quite agreed with my idea, and said that for 50 pound he could do his entire district. I hope he will bring it before the notice of the Government.

At Hanlon's Hotel, from which I took this sketch, I paid the biggest price for drinks, 2/3 for two lemonades and whisky. I had never paid more than one shilling a drink.

At 2.30 I saw Carpenter, the dog, and Blackburn off, the former returning to Cue, the latter to the "Rock of Ages". I presented Blackburn with the order of the "umbrella" — in other words, I gave him my white one, lined with green, in remembrance of our pleasant meetings.

Whilst standing at the door of the Hotel the previous evening, a brake and four drove up, the leader of the party was a Mr Calvert, on his road up to Cue.

A.J. Calvert, M.E. Company promoter and owner of race horses, bankrupt 1895.

It was 3.45 when I mounted the box seat of the Royal Mail Coach "Star of the West", the only other passenger being a young woman going to Perth. The first part of the road was a new track cut through the bush, only just enough space for the coach to pass, and in some parts we had to descend & ascend some very deep dykes, but

Sammy Marsh was a good whip and had his five horses well in hand, but the jolting was something awful. It would not, however, have been so bad had the coach been full up.

Our first stoppage was at Wownaminya – 16 miles — a sheep station where we arrived at 6.15, and had a good supper of liver and bacon with the Squatter, his wife, who I forgot to mention travelled with us from Yalgoo, and the rest of the family, including the shepherds. I had a good room to sleep in. True the floor was beaten earth, but it was a good bed and clean sheets, the first I had slept between since I left the Island Hotel 9 days before. I enjoyed the change.

At Morris's Station I had had a chat with a Frenchman who was taking some horses across the country. I met him again here, and he amused me before going to bed, by telling me his history. It had been an eventful one. I should imagine he had left his country for his country's good.

I have made a sketch of the house, in the foreground is an Indian woman smoking a pipe.

Wednesday, December 4th. Up at five, lovely morning, saw at least 30 green parrots at the water trough. At this station the charge was not excessive — Supper, bed and breakfast 4/-, the first time I had found anything cheap in W.A.

We left at 7.40, passed two wild turkeys within 30 feet — fine birds, should have enjoyed one cooked.

Stopped at Ederga — about 25 miles — and inspected the "Good Friday Mine" belonging to a Mr French. Arrived at Gullewa at noon, twelve miles from Ederga. Here I inspected the following mines: the "Christmas Gift", the "May Flower" , the "Osborne", the "Gullewa King", and the "Reward". Whoever discovers a new field is presented with a mine which is called the "Reward".

We had a good dinner, and several drinks at 6d. each — the price of drinks had dropped from 1/- to 6d. We started off again at 2.40, passing through a sandy country with gum trees growing all around. 4.25 picked some leaves of the flannel plant — very interesting, exactly like flannel. 4.45 passed curious granite outcrop consisting of enormous stones, some of them perfectly round.

Curious Granite Blow.

We drove this day at least 62 miles, probably more, with the same team of horses.

5.55 stopped for a spell at "Pollards Well". Whilst watering our horses, a teamster's horses ran away, it was exciting, but no damage was done. Soon after leaving the well we passed a beautiful green spot, fine gum trees, and other trees and shrubs. This well was 15 miles from Gullewa — it was 8.40 when we arrived at Moriney — the place where we were to stop for the night, the last two hours being driven in the dark.

Fortunately the driver knew the road, so did the five horses, they had done practically without stopping 25 miles, the only stoppage being two minutes at Pollards Well.

Wownaminya to Ederga 25. Ederga to Gullewa 12. Gullewa to Pollards Well 15. Pollards Well to Moriney 10 (or Undanooka).

The place we stopped at was simply a galvanized iron shed, divided into four compartments, there were no windows. We had to knock the people up; it turned out there was only one young woman, the wife of the owner, and her baby, in the shanty. She told us that when it began to get dark, feeling nervous, she went to bed. It certainly appeared to be a most lonely spot, not a house or a creature near. However, she soon lit the fire and put the kettle on, so that in a short time we had some tinned stuff and tea. I had a good bed, for the sort of place, and slept well.

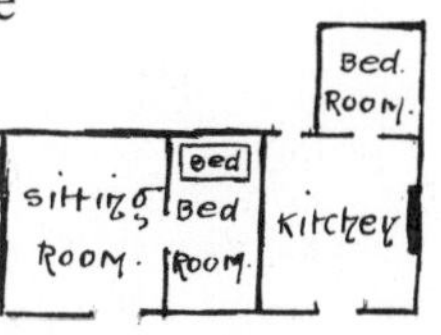

Thursday December 5th. Turned out at 5, a fine morning, breakfasted off kidney, bacon and onion stew, an Irish dish I should say, the young woman being of that nationality. We left at 8.15. At 9.30 saw a flock of black cockatoos with crimson tails, beautiful birds, very rare, and cannot be got to live in captivity.

11.38 arrived at Malachi Station, we had then come 20 miles. The horses were taken out, and we had our lunch, consisting of tinned lobster and pineapple, washed down with bottled beer, which I had purchased off the Irish woman, although she had no license to sell beer. This station was apparently deserted, the house was locked up, and we could see no sign of life about.

We left Malachi at 2 p.m., and drove into Mingenew at 5.40 — another 20 miles. So that the five horses did 102 miles in two days. This was too much for the horses, but some mistake had been made about a relay, and Sammy Marsh was obliged to push through with the one team.

Some parts of the country which we passed through were very fine, and beautifully undulating; in the distance, to the right, forming as it were a background, was the mountain range that runs by the Irwin River. The trees were of fair age — principally gum, and wild flowers were most beautiful. With the exception of the kangaroos, emus, wild turkeys, parrots, hawks and a few small birds, we saw no other animals or reptiles.

The road into Mingenew was dreadfully heavy for the poor tired horses, it was like driving on the sandy shore. By the side of the road ran a deep gorge with big trees growing up the slopes; at the bottom of this gorge should run a rapid torrent, tumbling in its course over rocks and boulders, no doubt. I have pictured what it is in the rainy season. At the Hotel — not at all a bad one — I had a good supper — chops and onions. They seem gone on onions in these parts, and plenty of excellent bottled beer. I was so pleased at having arrived safely, that I invited all at the table to take beer with me, which they promptly did. After supper I conversed with the Landlady, who had had 15 children — all alive — and at eleven o'clock, closing time, walked up to the station, and slept on a wooden bench until the train came in at 1.30.

Friday December 6th. Left Mingenew at 1.30, Watheroo 5.30, had breakfast, saw some more of the trees with the lovely golden orange bloom on the top of the leaves. It is called the "flame or fire tree".

tree with bright yellow bloom on the top of the branches. called the flame tree.

Moora 6.45, Mogumber 7.50, Mooliabene 8.40, Gingin 9. A very pretty village, the river Moore running through it, spanned by a timber bridge — Muchea 9.35. Twenty five miles from Perth the Darling Range of hills commences, and continues as far as Bunbury, which is on the sea coast. The range extends about 160 miles. Midland Junction 10.38, Perth 11.30.

Distances travelled by Railway, Coach and Buggy.

Railway	Perth to Mullewa	384	
	Mingenew to Perth	234	618
Coach	Mullewa to Cue	254	
	Yalgoo to Mingenew	118	372
Buggy	Whilst at Cue	16	
	Cue to Yowerragabbie	86	
	Yowerragabbie to Pinyalling	80	
	Searching for soak	10	
	" " road to B. R. Mine	15	
	From soak to mine	15	
	Pinyalling to Yalgoo	89	
	Searching for road to Yalgoo	21	332
18 days travelling	Total mileage		1322

In 18 days I had travelled 704 miles by coach and

Perth from the Mounts Bay Road foreshore.

buggy, and 618 miles by railway — at not more than 20 miles an hour, generally much less. It was a very trying time, but I came out of it, if possible, in better health than when I started. Everyone I met said, how well you look. I was brown or red, I was a deep bronze tint, and my hands a deeper shade.

Directly I arrived in Perth, after leaving my things at the Hotel, I walked direct to the Oyster shop. I consumed 3 dozen oysters, and a plate of fine red shrimps, washed down with stout — I did enjoy it, you may be certain.

On my way back to the Hotel, I met Mr Horgan — the first thing he said was — How's your Son? What's the matter with him? — said I. Says he, "blood poisoning". The devil, said I. Immediately after lunch I took the train to Fremantle, and found him looking the picture of health. He had been ill for a day, but soon got well again.

I returned to Perth at 9 o'clock and had a real good night's rest.

Saturday December 7th. Bobby returned from Fremantle. Interviewed Mr Horgan on various matters, packed our portmanteau, and caught the 3.30 train for Coolgardie. So I had not had much repose after my Cue journey.

It was 6.30 a.m. when we arrived at Southern Cross, 248 miles. Being Sunday there were no trains, so we passed a quiet day at the Cross; in the morning we went to the "Fraser South Mine", a fairly good one. Capt. Hill, the manager, gave us a good deal of information about the Mine, and presented us with a specimen. When going to this mine we passed the shanty where "Deeming, the murderer" lived. He was arrested at Southern Cross, where he was working as an engineer.

Deeming the murderer was mining engineer to the Fraser Mine, Southern Cross. When arrested he was dining at a so called hotel, kept by "Gerald" (Pearson) Frank Gerald – actor, etc. A very few days before his arrest he had mentioned at dinner table that he was engaged to Miss Rounsvellw who would shortly be his wife. He had, it appears, prepared for her: "the barrel of cement was there ready for her interment".

Monday December 9th. Left by the 8.30 train for Woolgangie — 74 miles. Bird was there to meet us with "Lulu" and "Kitty", and we drove to Bullabulling — 23 miles, where we slept at the Hotel which was kept by a man named Palmer.

Tuesday December 10th. Left Bullabulling at 6 a.m. for the "First Find". arrived at 7 and left at 10.30 for Coolgardie, getting there at 1.15 in time for dinner. After dinner I had conferences with our Solicitor and the Public Prosecutor about Lowden's case, which was on the following day.

They were quite pleased to see us again at the Royal Hotel, and regretted not being able to give us one of the best rooms, but since then we have the best in the house.

Six of the jury were in favour of acquittal, 5 being Jews, and 1 a German. When the foreman was pointing out to the German – he was a jeweller – that Lowden had spent about 6 to 7 hundreds on jewellery, he exclaimed "Mein Gott, I must be getting slow, that I did not strike up against that man." Another, a Publican, said "If a man spends 100 pound in my house I don't care if it's his money or his Company's." A third said "If my son was in the position of the prisoner I shouldn't like to see him convicted." The above was told me by a Mr Read, one of the Jury. About 4 years after, he was working as a labourer in Fremantle.

Wednesday December 11th. Lowden's case came on at 10.30. It was proved up to the hilt that he had robbed us of over 1000 pound, but the Jury disagreed. The Warden sent for me whilst the Jury were considering their verdict. He said — I wish to study your convenience Mr Tyler, I don't think the Jury will agree to a verdict, in the event of them disagreeing, do you desire to continue the prosecution. My reply was — certainly not — I consider it has been proved that he robbed the Company, and I have no desire to remain in the Country another two or three months, and then possibly the Jury disagree again. So when the Jury returned into the Court, and the Foreman said they were unable to agree, the Warden said the prosecution would be withdrawn, so the thief left the dock once more a free man, but every one knew he had robbed us, and we have the sympathy of all the respectable inhabitants in the City. He has now shaved his moustache, with the view I suppose to altering his appearance. Really, I am not sorry that he was not convicted, although he is an utter ruffian and deserved it, if any man did.

Thursday December 12th. Started with Mr Barnier and Bobby at 5 p. m. for Hannans — 24 miles — arrived in time for breakfast — hired another buggy, and drove to inspect the "Boulder West". Met Tebina

Lane, the manager of the "Boulder", driving into Hannans, pulled up, was introduced to him and had a conversation about the mine I was going to see.

Mr Reg B. Pell.
The Foreman of the jury, Lowden's trial, and keeper of the livery stables.

Returning, called at "Hannans Brown Hill" mine, and had an interview with Mr Varden, the manager, who allowed us to inspect the mine. We were put into suitable garments, and conducted through the workings. It is a very interesting mine, and will probably prove to be the richest in the Coolgardie or Hannans districts. The machinery was particularly well erected, but it is rather experimental, being a mixture of Dry Crushing, and Cyanide. It has only been tried once before — the ore is taken to a height of 40 to 50 feet, and there crushed, then by means of chutes is passed into a revolving sieve, then into another at a lower level, from whence it passes into a circular vat, and is treated by cyanide. If the machinery is *right*, but I am rather doubtful, the mine will be.

The ore is very rich, and we were presented with a specimen before we left. The machinery is to be worked by four Hornsby Ackroyds oil engines. I was sorry to hear that they sold specimens from this mine, it is a bad practice, as they would probably be bought to assist in floating some worthless properties.

With Mr Varden, I met Mr Bainbridge Seymour, a well known mining engineer on the fields. We arrived back at Hannans in time for dinner; we then had various drinks with various people, amongst

Bainbridge Seymour died in London about two years after my meeting with him.

others with Tebina Lane, afterwards we strolled to the top of the one street where there was not a mine, sat on a heap of stones, and talked of things in general. I went to sleep leaning on another stone — it was hard — but when you are tired you can sleep anywhere, and any how. At 8.30 we turned in, and both slept well.

Friday December 13th. Had an interview with a Mr Jephson, formerly Warden of Hannans, and a Mr Tom Boner about a panorama illustrating W.A. scenery, mines etc. etc. After dinner at 2.30, started for Coolgardie, had a beautiful drive back by the old track, much more picturesque and interesting. In the middle of the journey, we found that the water bag was missing. We drove back and found that a carter had picked it up — he not only returned it, but put some water into it. They don't give much water away in this country, so I looked upon him as a generous person. It was 6.40 before we landed at the Hotel; we had been transferred to the best room in the house, very comfortable, no possibility of the sun shining in, so cool and pleasant.

Saturday December 14th. Very busy all day, sending cables, writing letters etc. etc.

Sunday December 15th. Went to the Bank with Bobby to see the gold made up for transport to Perth; a very interesting sight, the care taken in weighing it is surprising. I purchased a few more small pieces. We lunched and dined with Mr Charles Kaufman — a mining engineer of first class position — at his Bungalow; Mr Barnier was with us. We enjoyed the day greatly — it was more like an English home, well furnished and most comfortable. Mr Kaufman is a great authority on mining matters, and I gained considerable information from him. During the evening there was a most terrific hurricane — it blew frightfully, and the dust was awful, but no rain fell. It was a "dust storm".

Mr Kaufman's opinion was, that eventually all the mines would drop to one to two oz shows. Would not at present go outside a radius of 50 miles from Coolgardie.

Monday December 16th. I have been engaged on this letter all day. I was determined to let you know all I had been doing, as I know you are interested in my movements.

With Love to you and Lulu,
Your affectionate Husband,
Robert Emeric Tyler.

Dec. 18. Interview with Dr Davy. In his opinion the following mines are good: Lady Lock — Lady Forrest — Lady Shenton — Wedderburn — Golden Arrow — Black Flag Proprietary — Hannans Brown Hill — Great Boulder — Lake View — Ivanhoe (moderate) — Golden Horseshoe — Hill End — Burbanks — Hit or Miss — Carbine — New Victoria — Australasian Red Leap — Empress of Coolgardie.

The Cleopatra Hotel
Fremantle,
Thursday, Nov. 21st 1895

My Dearest Mater,

This time you are going to have a long letter, as I did not write last week. The Pater wrote you up fully about our journey from Coolgardie. It was about the worst time I have been through at present — I was not at all well all the journey so you can fancy being jolted about, your eyes full of dust, and a fearful headache — but when we got to Perth the hot bath, and oysters, compensated for it thoroughly. We did a good bit of rowing last week, in fact my hands are in such a fearful state that I have had to leave it off for a spell.

On Monday I went for a bathe — it was rough (at least the Boatman said so), the Westralians are an exceedingly cautious lot. I got him to row me out, and then plunged in and swam around, it was simply grand. On Tuesday night we went to a very jolly evening at a Mrs Milward's (a lady I had met on board the steamer). The two Orchards were there, and we regularly enjoyed ourselves. On Monday night I was bitten for the first time by mosquitoes, and ever since my hands are all over small bumps, still they are not so bad, they irritate a little, but not much. Then on Friday we sent off the Jewellery and also my specimen box, which I hope will not be

The Cleopatra Hotel, Fremantle.

opened till I return, there is nothing of interest unless first explained.

On Saturday morning the Pater and I started for an excursion. We took some sandwiches, and whisky and water, and chartered a boat, I rowing, the Pater steering. Well, we pulled on for about a mile, and stuck on a sandbank, so we had to row all the way back again and start afresh, after about 3 miles of it I began to get a bit fagged (it was not like river pulling) so Pater took one of the oars, and pulled manfully for about two miles, as we had the wind and waves in our teeth. This was on the Perth Waters. At last we passed under the Canning Bridge, and were on the Swan River where it was much calmer, we then

Excursion on Perth Waters.*

[* The text description suggests in fact that they travelled west through the Narrows and out of Perth Waters, and then south on the Swan River towards Melville Waters, then passing under Canning Bridge and into the Canning River.]

moored the boat to the bank, and the Pater went collecting some of the "Flora" whilst I paddled about. Then we had a try at fishing, but having no hooks we didn't catch much. Returning we sailed the bark very successfully. We arrived home about 5.30, after having had a glorious day.

On Sunday, the two Orchards went with us, and we had another fine day. On Monday we messed about buying things for the Pater — can you imagine him in a yellow suit of Khaki, a soft shirt, a big solar topee and with a net coming all over his face, and white umbrella lined with green, besides a water bottle slung across his shoulders — he looked just like an Indian Major. He left for Cue at 6 o'clock. I should awfully have liked to have gone with him, but it was too expensive, it would have cost getting on for 50 pounds for the fortnight, so I left him starting for Cue, two days in the train, and three days on a coach.

Then I came down here, and am staying with Whitworth, the young engineer, and am having a fine time. We get up at 7 and go bathing in a small creek (of course here it is the real sea). There is a reef about 40 yards from the shore — you can swim out to it but no further as on the other side there are sharks dodging about. Still, it is very nice swimming out there and sitting on the reef (the water is about a foot over it) and diving off it.

On Tuesday the mosquitoes were much worse, big lumps as large as 3d. bits all over my hands. On Saturday we are going to have a rare treat. There is an island called Garden Island about 10 miles off the coast, quite uninhabited, with plenty of Wallabi, opossums, snakes and parrots, and in the sea huge schnapper are to be caught (they are the fish of the country, sometimes weighing 60 to 70 lbs). We shall start on Saturday morning, sail over there with some tucker, two guns, some rods and rugs, and camp there Saturday night, and on Sunday try what sport we can have. If we do bring down a Wallabi, Whitworth is going to skin it for me, and I shall bring it home as a curio; this will also apply to opossums and parrots. It will be like Robinson Crusoe.

Feb 17. 1905. The Australian Liner the Orizaba, 6,300 tons, went aground off Garden Is. Owned by the Pacific Steam Navigation Co., chartered by the Orient Co. Passengers and mails landed at Fremantle.

I shall come home a jolly sight stronger than when I left England — I can tell the difference now. I expect we shall spend Christmas either at Adelaide or Melbourne; it will be beastly dull, but still we shall be in something like civilization. The next letter, or the one after that, will be interesting as it will be the Cue letter from the Pater. It will be funny to wear a high hat when we get home, really the Colonial way of dressing and wearing soft hats is much more comfortable.

The Pater is looking magnificent, as brown as a berry, and in fine spirits and health. I hope you will like the small presents we sent — we thought we had better mark them or there might have been a fight.

It is much cooler down here, and you get some splendid sea breezes. This is a Commercial Hotel, so I know many of the Travellers of the big firms. Tell Percy I have jotted down a few of the advertisements I have noticed, but the majority of them are not worth looking at; they will be better when we get over the other side.

I shall become a regular sportsman — I have already learnt to ride, to drive and to row, and may in course of time learn to shoot and sail, then I think that is about all. I am quite a big personage here, having just come down from the fields. Bird is left up at Coolgardie, I don't think he likes it a bit. Look out for the S.S. *Rome* as my box is on board her, it is insured for £100 so I don't know whether I want it to be lost or not, still I think it is worth more than that. The Pater is still collecting up at Murchison, so I shall have another box. The people at the Grand Hotel Perth were quite delighted to see us again. Hoping every one is in the best of health & spirits,

I am, Your loving Son,

Bob.

The Fremantle–North Fremantle Bridge, 1907.

Cleopatra Hotel,
Fremantle Nov. 29th 1895.

Dear Mater,

As you see by the address, I have not gone yet which means that the Pater is still wandering about the wilds. There is one good thing I don't think he is alone, as Mr Carpenter is with him. I had one telegram saying he had arrived safely at Cue, and then another that he was going to Yalgoo, and would then return here on the 4th. By the time this reaches you I suppose it will be after Christmas — did you like the cards and jewellery we sent you? I often wonder if my box will arrive safely.

I think in my last letter I mentioned the trip I was going to take to "Garden Island" — we did intend to get up a party, but were disappointed at the last minute, so Whitworth and I had to go alone. Our chief difficulty was in finding a certain boatman; we were particularly anxious to get him, as he has the reputation of being the best boatman about here.

On Saturday at 1 o'clock p.m., just as we were giving up in despair, we ran across him. He agreed to go, so we all three started off to get a boat. After a little difficulty, it being late, we secured one about three tons, carrying a mainsail, a staysail and a jib, partially decked over, capable of carrying 7 or 8 persons, so there being only three of us we had plenty of room. After arranging all about the boat, we agreed to meet at 3.30. Whitworth and I came back here, had a little dinner, then went to purchase the tucker, as follows — 1 tin of tongue — 1 tin of corned beef — 1 tin of tomatoes — 1 tin of plum-pudding — 1 tin of jam, 1/2lb of tea — 2 tins of Swiss milk — 1 pepperette and salt — 4 loaves of bread, a large canister of water and a bottle of whisky in case of accidents. At 3 we started taking with us a change of clothing — also 1 Colts Magazine rifle, and a shotgun, 4 fishing lines and a camera. Of course I am not mentioning the billy, plates etc.

At about 4 we started, from the time we waded off to the boat until returning we had not our shoes and stockings, or rather socks — on.

It was not a good day to go, it being rather rough, and overclouded, and the wind not in our favour. The distance to Garden Island is about 11 miles, and all the way there are reefs, so you have to keep in 7 feet channels, and sometimes narrower than that, in fact practically speaking, you can see the bottom the whole way. As we got further out, it became rougher, and we got occasional duckings. A little bit nearer the land — about a mile — is a much smaller island called "Carnac Island". When we got near it, it was about 7, just light enough to see. Eddie (the boatman) decided not to attempt to get to Garden Island that night, it being too rough, so we ran into a small bay on Carnac and grounded. I jumped over and waded to the shore, rather, if the truth must be told, glad to get on terra firma again. Still, contrary to my expectations, I had enjoyed the run over immensely.

Fremantle Jetty.

Eddie saw to the sails and Whitworth to the anchor, then we carried our tucker and our bags, rugs etc., on the sand, and went prospecting for a good site for our camp. We were damp, to say the least of it, and to add to it, it came on to rain, a sort of sea mist. We selected a site, a little sheltered by some bushes, then it became quite dark and unfortunately we had forgotten to bring a light with us. Eddie was taking down the main sail, W. was fossicking about for firewood, and I was exporting the things from the shore to the camp. Then, after some difficulty, we got up the sail to form a screen against the wind — by this time we had got a good fire burning which gave us a light; we had only one rug, mine being stolen at

Coolgardie, which we laid out on the sand, and commenced getting our tucker ready.

You can imagine after all this exertion, being damp, what a Godsend a cup of tea was — all of a sudden a cry rent the air — on looking up we saw W. making a grimace. We wanted to know what was up, he replied "no sugar". Well, the air went blue for a little time, but as there was no use in crying over spilt milk, or forgotten sugar, we made the best of it by putting in lots of Swiss milk. Then another exclamation was heard — this time it was myself — we had left the corned beef, together with tongue, in one of the lockers, and as it meant swimming out to the boat, which none of us relished, it being then about 10 o'clock, we made the best of it, and fed off bread and butter and tomatoes, with fig jam to follow. It was drizzling all the time, but we kept up our spirits wonderfully well so after clearing up and piling up the fire, we turned in. We followed the tip of the Carthaginians when in Spain — i.e., digging away the sand where our hips and shoulders come and heaping it up for a pillow. After a stiff "half-un", which I think prevented us taking cold, we told a few yarns and then dropped off to sleep. As it was rather novel I did not sleep well, as you can imagine, at least I don't think you could imagine the weird outline of the bushes, the lapping of the waves, and the deadly stillness of it all, the magnificent stars overhead – (I found myself humming Lulu's song — "Stars"), and the chance of seeing a snake wriggling past — well about 4, just after day break, I was feeling a bit cold, so I arose, shook myself and started out, leaving the others snoring. I walked down to the sea, then climbed over some rocks and on to some hills, got a few curiosities and then after about an hour, I saw that Eddie was up, so I slowly returned, dragging a hunk of wood with me for firewood. Then W. went off with his gun to see what he could discover, and I stripped and swam out to the yacht, got on board, and pulled her in, then went for a short swim, returning feeling I could do anything. W. returned having seen some rabbits; he wounded one, but it got away into its hole before he could catch it. Then he had a dip; after that we had our breakfast, we did punish that beef for hiding itself away the night before.

About six we started, it was a better day, but still rough. Oh! I forgot — we took a snap shot at our camp which, if it turns out well, I will either bring or send home. It was beautiful. At 6.30 the sun came out and then we sighted Garden Island. Coasting there we saw some shags — a large bird about the size of a goose — they feed on fish. Well, W. shot 3 of them, and it was great fun picking them up as we sailed past; unfortunately it was too rough to fish, so we lost some good fun as the fish here are the size of large cods — schnapper — it is fine sport hauling them in.

The sailing was lovely; about 12 we got to Careening Bay, where we waded ashore and boiled the billy, having our dinner on board, as it was about 100 yards to the shore, the water up to your knees, so it would have been rather a fag to cart all our tucker to the land, and then what remained back again. Whilst the billy was boiling W. and I had another bathe, and I can tell you we did lay into the grub afterwards. Then when Eddie was doing the washing up we fished, catching about 30 small Trumpeters — about the size of our perch, and after that a few herrings, all in the space of a quarter of an hour. Then another yacht came up, and after a talk we proposed a race to the "Dummy" (a patch of water about 20 yards in diameter which is always smooth, even in the roughest weather; no one can account for it, it is a long way from land and altogether very curious). You can imagine, it was difficult to steer to it, the distance being about 4 miles. The other was a faster yacht, but we were lighter; it was a grand race, but the superior knowledge of Eddie came in, and we ran in first, winning easily.

We then had another shot at fishing, but it was no good, after that we started for home, when we discovered that we had left one gun and our shoes at Careening Bay (intending to go inland to have a shot at some Wallabi). So we had to return and pick them up; still all's well that ends well, we arrived home about 7, Sunday evening, as brown as berries, and dirty. We had a good tea and went to bed at 10 o'clock, sleeping till 8.30 the next morning, having had the finest time I ever remember. I am becoming a regular bushman now, and rather like the life. All the week we have rowed about, bathing every morning and in fact getting as strong as horses — I met a man I knew up at Coolgardie, and even in this short time he said how much better I looked, stouter etc.

View north across the Swan River to North Fremantle.

Yesterday was a great day, the first Admiral that has ever been to W.A. arrived yesterday — H.M.S. *Orlando*, Admiral Sir K. Cyprian T. Brydges: well, you can guess the excitement, it was tremendous, they had all the W.A. regulars and volunteers out to meet him, as a guard of honour (yeomen) and a salute of "two" guns. Of course, as you know, whenever there is something special such as a holiday or a fair, the people get drunk, so the fun in the evening was great. Whitworth is a jolly fellow, about 22 and rather like young Spen; we get on very well together. Now how are you all? By jove what would I not have given to have been with you all at Christmas. I suppose the family dined with you.

Your loving Son,
Bob.

Cleopatra Hotel,
Fremantle, Dec. 5th 1895

Dear Mater,

The Pater, I think, returns tomorrow, and then has to start away immediately to Coolgardie for this blessed trial. I don't know whether I shall go with him, but I hope so. I have not been doing much this week except that I went to Garden Island again — this time I started at 11 at night. During the week we had got up a party of seven with ourselves and the boatman (Eddie Brown), eight souls in all. We got a larger boat, and all the provisions, and as some of them could not get away from business, we arranged to start at 10 o'clock. We had all ready on the beach, and waited until 11 o'clock, not one of the fellows turned up. It was beastly mean after they had all promised, you can imagine the difference in expense dividing by 2 instead of by 7. But there was no help for it so we started.

It was very rough and after we had got well out, Whitworth was ill. By a most wonderful coincidence I was all right, and as there was a good wind we ran over there in an hour and a half. I steered all the way, one or two seas came over us, but not much. We anchored and all slept on board (it being a larger boat than we had last week). I did not sleep much and saw the dawn about 3, it was really beautiful. Then the others woke up, at least Eddie did, W. did not feel well, so slept on until about 10.

We then sailed to the "Dummy" (which I mentioned in my last letter) but unfortunately it was again too rough for schnapper fishing, so we turned back and ran to Garden Island again, having breakfast there. We had two shotguns on board, and a rifle, so I had two or three long shots at shags. After breakfast, W. feeling better, we decided to go to Carnac Island to have a shot at the rabbits. On the way I shot a shag, a fine big fellow, and W. 3 gulls. When we landed W. shot a rabbit, I collected a lot of curios; we then had a bathe and started for home, sailing round H.M.S. *Orlando* as we returned, arriving about 7 very sun burnt and went to bed at 10.

That night I was not at all well, but got all right on Tuesday morning — we have done nothing all the week, except watch the ships unload on the jetty. I am going back to Perth on Saturday expecting the Pater to be in about 6 or 7 in the evening; he will have a lot to tell me. How are you getting on? Anything fresh happened? I keep looking forward to when we arrive. What a fearful amount of excitement there will be. Of course the Pater has written to you.

With best love, I remain your loving Son,

Bob

Coolgardie,
Sunday, December 22nd 1895.

December 22. 112° in the shade.

My dear Emma,

This week I have very little to say, I exhausted all my interesting news last week. I really can only write about the heat; for the last two days it has been 112° in the shade — fancy that. It is now 5 p.m. and it is 100° in the office where I am writing. My hands and arms are covered with beads of perspiration, it is running off me, and where my arms rest on the table, are pools of water. At lunch — 1 p.m. — it was simply awful, the perspiration seemed literally to jump out of you. You can't eat much, well! It would not be good for you, but you are obliged to drink; for lunch I generally take beer, but having a slight pain in the back — which I put down to liver — I gave it up for a few days, but am now all right again.

The drink lately has been Hock or Chablis, with soda and ice, a good refreshing non-intoxicating drink. Ten shillings a bottle — quite enough — the latter is a beautiful wine, a product of the country. I shall endeavour to get some for you when I return to England, as it is just what you will like.

December 19.

I have been alone since Thursday. Bobby went to the new mine, and to prospect over the "Lulu", the property adjoining. I hope he will find that there is some good gold.

Our Manager found some prospects there the

other day. I am now expecting him back to dinner at 6 p.m., but owing to the excessive heat he may be later. I am awfully pleased that my journey is drawing to a close, but it has been a "great experience" and although I have had many hardships, still I've not had a bad time. But to think I am spending my Christmas away from you and Lulu, and all relations and friends, seems passing strange. I do hope you will all enjoy yourselves notwithstanding my absence and Bobby's, we shall drink your health, Lulu's and the rest of the family in champagne, and shall hope you have done the same.

All the best people have left, or are leaving tomorrow, for Perth, to pass Christmas. Unfortunately I cannot get away as I have to arrange for a new Manager, and on Tuesday, I am taking an Expert to the South Londonderry. At 7 Bobby returned, had enjoyed the change at the Mine. He drove "Lulu" & "Kitty" back; he always drives now, and is becoming a first class whip.

December 23.

Yesterday, Monday, we drove with a Mr Breakspeare to see three mines near Coolgardie, specimens of which we brought away to add to our collection.

The Queensland Development and the Old Chums, passed by the Big Blow, when going to the former.

Today Bobby has driven Capt. Vaudrey to the South Londonderry Mine. He is to advise whether we should continue working it. I did not feel very well (liver) so sent him instead. Dr Davy is looking after me, I'm much better now. The shop people are decorating the fronts of their shops with green trees and branches, but it is not like Christmas. It is only

82° in the office — quite cool — well, this morning the water was absolutely cold; when we went to bed last night it was "hot".

Give my love to Every one,
Your affectionate Husband,
Robert Emeric Tyler.

Christmas at the White Feather Cement Claims.

Coolgardie,
December 30th 1895

My dear Emma,

December 25.

The sun rose bright and brilliant o'er the Street, and the other, as yet undeveloped streets of Coolgardie, the coming city of Western Australia. It was Christmas morn. The one cow could be heard lowing in the distance, the plaintive note of the domestic hen, coupled with the more robust note of the ancient rooster, alone broke the stillness of the early dawn.

The dwellers in this would be golden City of the West, had decorated their door posts, and lamp posts with young trees, and boughs cut from the adjacent bush. All was green — but the "Dwellers". Even the few melancholy gum trees — left standing in the street to remind future generations, after all had fallen under the ruthless axe of the Miner, that gum trees once flourished — looked green and shady (12.15 p.m. 95° in the office).

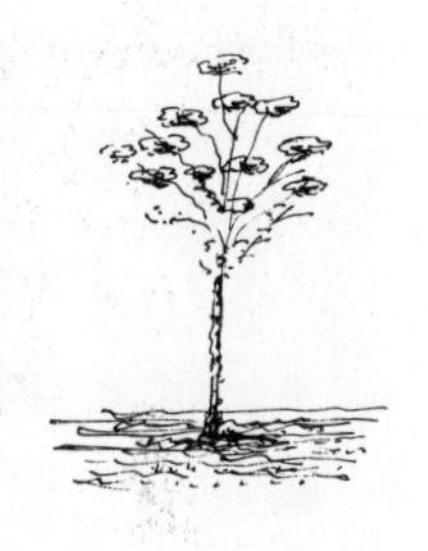

The sun rose higher in the cloudless sky — out from the highways and byways stagger swarthy miners tanned by the burning heat of many a day passed on the brace under the sun's cruel way — (the brace is the top of a shaft). Far away in the waterless bush, where animals live not, nor birds find roost, they struggle and toil, unearthing the soil in the search for gold.

It was impossible for one single moment to imagine it was Christmas — now Midsummer day, you might have thought it, the heat was awful, but at the Hotel, fortunately, they had plenty of ice, so we were able to get cool drinks.

Christmas Dinner.

The dinner was in the middle of the day, the same as on Sundays. Mr Barnier and Dr Davy dined with us, we had Chablis and Champagne, and we drank in the latter to absent friends, Wives, sisters, brothers, daughters, nephews, nieces and cousins. We had goose, chickens and plum pudding, but no turkey, of course, soup, roast beef and other good things, in fact it was not at all a bad dinner, but boiled fowls satisfied one; I was not quite up to the eating part of the business. After dinner we sat on the balcony and smoked and talked, then went to

Christmas Dinner.

sleep until tea time. After tea I went across to the stable to order the horses to be ready for 6 the next morning, as I had arranged to drive Mr Barnier to the new Mine. All the horses, some twenty or thirty, were stabled under open sheets. I was unable to find "Lulu", but on speaking to one of the ostlers he pointed her out to me. She was in such a frightful state I failed to recognise her, covered with dust and dirt. I could see she had just come in from a journey. On making enquiries I ascertained that young Pell, the brother of the owner of the stables, had sent her out as one of the leaders in a coach. The party, who had been to the Londonderry Hotel, all got drunk, smashed one of the wheels, left the ladies five miles out in the bush, and drove back with a tree under the hind axletree, as hard as they could. I was furious at such a liberty having been taken, and the next morning removed the horses to another stable. I should mention that the drunken ruffians forgot all about the "ladies", and they had to walk into Coolgardie.

Boxing morning at 6.30, Mr Barnier, Bobby and I started for the new mine. Bobby rode on a small horse, a gentle horse, but somewhat wilful — colour cream. He objected to cantering, and Bobby to trotting, so did not agree as well as could have been desired. However, by dint of the boot, or rather both boots, and a stick, they got on fairly well together, but I fancy they both were equally well pleased when the journey was at an end.

3 p.m. 101° in the office.

Bobby had discovered a new track through the bush, so was our guide on the occasion; this track had one great advantage, we avoided the dust. We had absolute faith in our guide, leaving ourselves unreservedly in his hands, and the "cream one". They ambled on, regardless of stumps of trees and other pit falls — we followed — until at last we noticed a look of concern and uncertainty on our guide's face. We had then been driving about an hour and a half; on questioning him, we found he was doubtful as to our position, in fact, on considering the point, thoughtfully, carefully, he came to the conclusion that we had missed the spot where we should have gone off to the right.

Confidence in our guide was gone. Where were we? And what were we to do? Without going back some miles, we were bushed. But not lost. Fortunately the gentleman with us was used to bush life; he noted the position of the sun, then the shadows cast by the trees and shrubs, and, turning the horses heads to the right, struck straight through the bush, driving over all obstacles, hills, bushes, young trees, and in fact everything that came in our way except big trees, and in some parts the bush was pretty dense. After a time we struck the track we should have been on, and once more we were happy. Fortunately we had a water bag, a bottle of whisky, and another of lime juice.

Interview with Mr Thompson, president of the Chamber of Mines, at the Club in the morn.

We arrived safely at the mine about 9.30, and enjoyed thoroughly some fresh eggs which Mr Barnier had brought with him, and some cold beef and bread, which we had also brought in the bag, which Louisa presented me with. Of course we had the usual tea — nothing can be done in this country without the "billy" — first boil your billy. I must confess that tea quenches your thirst better than anything else — cold tea and whisky with a little lemon is not bad. After inspecting the mine, and generally looking round, we had a pleasant drive back to Coolgardie. I fancy Bobby and the cream coloured one hit if off better on the return journey.

We have been much concerned at not receiving any news from you for the last two mails. I assume you have written to some other port. I imagined Mr Pugh would have kept you informed as to our movements. Our journey is now drawing to an end as far as Coolgardie is concerned. We hope to leave on Sunday or Monday, and glad we shall be to turn our backs on this place. It is too hot, although we are not so badly off as many others; in their offices it has been up to 112°. My next letter will be, I suppose, from Perth. I hope you all spent a happy Christmas, we are wondering if you passed it at home, or at Louisa's or where.

Your affectionate Husband,
Robert Emeric Tyler.

P.S. by Bobby. A happy new Year and prosperity to all — 1896. This is a most auspicious year, I reach my second decade on May 8th. Hurrah! We are leaving the land of sand, sun and sorrow on Sunday — oysters, baths etc., at Perth, and then for America.

P.S. January 1st 1896. It being New Year's Day, I

December 27. Bobby & Bird drove to S. Londonderry for chaff.

Received Capt. Vaudrey's report on the S. Londonderry.

Attended cricket match on the fly flat (no grass), Coolgardie bank clerks v. Hannans, and dinner in the evening. Returned thanks for the visitors.

Had egg-flip with Barnier. Met Jack Borleau, first chemist in Coolgardie. Took afternoon tea with Mrs Jerger at Toorak.

The correct saying is: "The land of flies, sun and sorrow, sore eyes and Mining Experts".

Interview with the new manager, John Freeman. Bobby with him to the South Londonderry — too ill to accompany them.

determined to start another page. Well! It is opening another page in one's lifes history. We know not what is in the future, it resembles an unread book, with the leaves uncut. When we look back at last year, the opening day of which we passed at St. Johns Wood, little did we anticipate what the year was to bring forth, and as to this year — well! If it brings one back safe and sound to England, I shall be satisfied.

December 31.

Yesterday morning we went to see the largest nugget as yet discovered in W.A. 303 oz. valued at about £1,111, gold being worth at Coolgardie about £3.16.00 an ounce. Sometimes when it is exceptionally good it is worth over £4. It was a beautiful chunk of gold, made your mouth water to look at it. The length was 10", depth at the deepest part 7" by about 2" thick.

The Joker

Nugget. 303½ oz - found at the Black Flag, on the Devon Consols Mine.

It was shown to us by the lucky man who discovered it, at one of the Banks, and very careful they were of it. We went with the two Miss Edwards and two or three others to inspect this interesting specimen of gold. The man who found it was there, a grey bearded old man, he told us he was working in a hole about 6 feet deep, when he came across this nugget and some smaller ones weighing about 300 ozs. That's what I call luck — in fact finding gold is all luck. I enclose a cutting from the Coolgardie Miner, which will satisfy those who have read my letters that to be bushed is sometimes a serious matter, poor fellows, they must have had a frightful time of it before they died. Last evening I saw Constable Brown who buried them, he looked as if he had had a bad time.

January 2. Interview with the new manager. Instructed him to carry out all the suggestions made by Capt. Vaudrey.

103° on the balcony of the hotel.

January 3. Bobby drove Freeman and Bird to the First Find. Too ill to go.

We are sending you a photo of the buggy & pair, also a view of the great fire. We imagined they had been burnt in the last fire, which occurred whilst we were away, but fortunately they were preserved (the photographic studio was burnt down). I am glad of it, as you will be able to see what the buggy and "Lulu"and "Kitty" were like, you also get a view of my hat, a sketch of which I gave you some time back. Lulu is the one nearest to you, a pretty horse, if we'd only got her in London, what a time we could have.

It is possible a Mr Wills and his wife may call upon you in a month or six weeks — I have asked them to do so — they left here this morning. Poor fellow he has had a bad time with fever etc. And now I must pull up — With fondest from your son and myself, and wishing you and Lulu and all relations and friends a happy new year.

Your affectionate Husband,
Robert Emeric Tyler.

Extract from the "Coolgardie Miner" — December 31st 1895.

"The fate of the two unfortunate men, Thomas Cantwell and Jerry O'Conlan, who were lost some time back on their way from Dundas to Coolgardie, has now been finally cleared up, and adds one more harrowing chapter to the story of the goldfields. Having just disposed of a mine,

Coogardie in 1897 — from the Fire Tower.

they were coming in with their two mates to draw the money, when they got off the track. In a very weak state they managed to strike the Sunday Soak, only to be disappointed in finding water, and the other two men went off in search. When they returned Cantwell and O'Conlan had disappeared — wandered off into the trackless waste. No hope was entertained from the first that they would be discovered, but Constable Brown and a black tracker went in search of them. The missing men were followed to a point some 50 miles west of Sunday Soak, and there their bodies were found. They had kept close together despite the delirium of thirst, and of what they must have suffered ere death came to their relief only those who have nearly perished themselves in the wilderness can form an adequate idea.

There were signs by the way which showed that they had several times laid down and rolled on the ground in their agony. In the end one of the poor fellows had crawled into a sand pit about 3 feet deep, made a last despairing effort to reach water by scraping the sand out with his hands, and so he died. He was found still in that position, and only about four yards off lay the dead body of his mate, lying on his back partially stripped of his clothing. On their wanderings they had passed within 150 yards of a supply of the precious fluid that might have saved their lives. The discovery took place on the day before Christmas Day, and as the bodies were then much decomposed, they were given immediate burial close to the spot where they had breathed their last."

Extract from a Coolgardie paper — November.

"Miners have had to leave Mt. Ida district for want of water, the soak having suddenly given out. A general stampede took place and some narrow escapes were experienced, one man subsisting for 25 hours without anything to drink. The whole track is covered with dead horses and camels, swags, and provisions, the latter having been thrown away in a mad rush

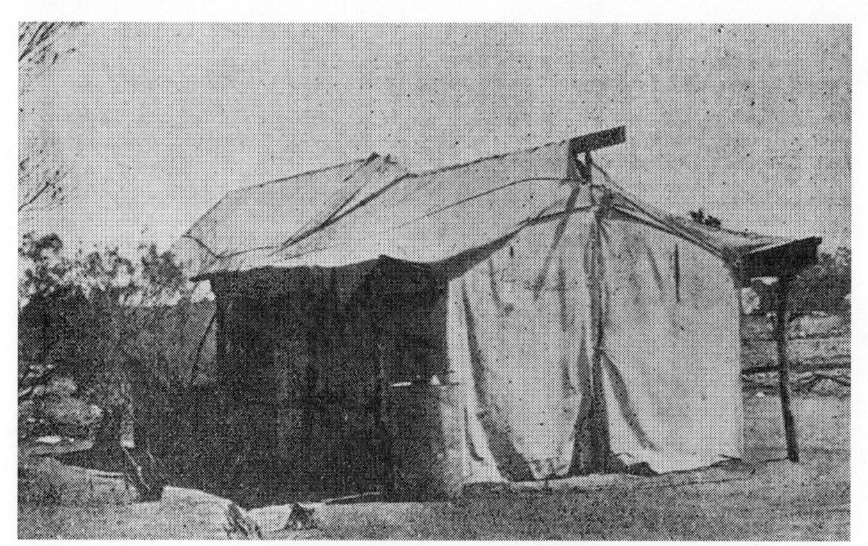

Old Roman Catholic Church, Menzies.

for water. The supply is none too plentiful at Menzies where the men have retreated. Ullaring has been deserted for similar reasons."

Extract from a Comic Journal — "A Most Distressing Case"

"I want some consecrated lye" he slowly announced as he entered the Chemist's.

Lye is a sort of soap made from ashes.

"You mean concentrated lye," suggested the Chemist, as he repressed a smile.

"Well maybe I do. It does nutmeg any difference. It's what I camphor anyhow. What does it sulphur?"

"Eighteen pence a can."

"Then you can give me a can."

"I never cinnamon who thought himself so witty as you do," said the Chemist, in a gingerly manner, feeling called upon to do a little punning himself.

"Well that's not bad either," laughed the customer with a syrup-titious glance. "I ammonia novice at the business, though I've soda good many puns that other punsters reaped the credit of. However I don't care a copper as far as I am concerned, though they ought to be handled without cloves till they wouldn't know what was the madder with them."

"Perhaps I shouldn't myrrh myrrh. We have had a pleasant time, and I shall caraway."

It was too much for the chemist. He collapsed.

January 5. Sunday. (Unwell.) Took Barnier for a drive — thermometer sunk 40°.

January 6. Purchase of First Field Gold Mine, Bulla Bulling, completed. Weather very cold.

Coolgardie
January 7th 1896.

My dear Emma,

Yesterday we received a cable to say you were all right in Gower Street. This relieved our minds, as we had not received any letters for at least three weeks, or more. It appears you've been writing to Colombo. I can't understand why Mr Pugh did not let you know our movements, he perfectly knew where we were. It is most annoying, naturally we are most anxious to hear how you are getting on in our absence, but thank the Lord the end has come, as far as this place is concerned, for we leave tomorrow morning by the coach, and awfully glad we are. The weather has been most trying, and it is wonderful how we've pulled through it — for more than a fortnight 112° on the balcony, on which our bedroom window looks, but on Monday morning it dropped to 72° and the following morning to 62. What do you think of that for a change — I had to get out of bed in the middle of the night to put my coat over me. When I went to bed I had to damp my towel to lie my head on, everything you touched was hot, even the water in the jug was as if it had been boiled, and just allowed to slightly cool down. You were simply walking about with your tongue hanging out, the only relief being iced drinks, but when the weather was hottest, the ice machine — the only one in the City — invariably came to grief.

Before we leave by the steamer, which may be in a week or ten days, we shall write again to let you know how we are coming back, but I think it most probable we shall go through the other Colonies and return by way of America. We may never have such a chance again, and see what good it will do your son, he will have been all round the world. At the most it won't make a difference of more than three weeks, and what an experience!! I suppose Lulu is saying, "What a time Bobby is having — when will my turn come?" Well it may come soon.

Now how have you been, and my dear daughter, I do hope you enjoy yourselves, and that the presents arrived safely, and that you all liked them.

I remain, Your affectionate Husband,
Robert Emeric Tyler

P.S. by Robert,

It is a jolly good thing we are leaving tomorrow as the Pater has not been at all well for the last 3 weeks, but the change will set him all right. You can bet we shall have a time when we return — only keep well and wait (it is rather sickening out here) and then enjoyment for all.

Ever your loving Son and Brother,
Bob.

Perth from Mount Eliza.

Grand Hotel, Perth.
January 15th 1896.

Dearest Mater & Lulu,

As you see by our getting to Perth we are a little nearer reaching home. For myself I have to give you very disappointing news, but I expect you will be glad.

The Pater had not been at all well up at Coolgardie, so we were very pleased to leave it for the last time; we arrived here on Thursday night. On Saturday night the Pater was taken ill, not dangerously, but very acute indigestion, excessively weakening. Today he is entirely free from pain, and I am taking him down to Albany on Friday to catch the

R.M.S. *Orient*, leaving there on the 25th. So you see all my hopes of visiting the Eastern Colonies and America are gone, still the Pater is regularly homesick, so we shall arrive in London about the 25th or 30th of February. You will probably have a telegram from Naples re the Paris trip. Things have been very quiet as you can imagine, everybody has been exceptionally kind to us, and on the whole we have got through the trip wonderfully well.

How are you all getting on? I saw Ingpen, he told me you were well and happy.

With best love to everyone, especially to yourselves,

I am your loving Son and Brother,

Robert.

Barrack Street, Perth.

Return Journey from Coolgardie to London

Wednesday January 8th 1896. The business that had detained us in Coolgardie was completed. The new mine — "The First Find", Bullabulling — was paid for, and all final arrangements had been made for the future working of the mine.

Having said goodbye to Mr Barnier and all other friends and acquaintances the day before, we rose at four, and at five started by the coach for Woolgangee, the point at which the Railway had then reached. Owing to the many drinks which accompanied the "good-byes", it was past 12 o'clock when we got to bed, so we were not particularly lively at the start.

Railway opened to Coolgardie on the 23rd March 1896 and to Menzies 22nd March 1898.

We had good seats on the coach, behind the driver, a lady and gentleman having the box seat. Our late defaulting manager came to see this lady off, but I guess he wished he had stayed away when he saw me seated behind her. As we drove off I wondered if ever I should see the City of Coolgardie again; we had had a rough time, but out of it, had found a considerable amount of enjoyment, coupled with a great deal of discomfort, but we had acquired much experience, which I hope may be turned to some good account in the future.

Bird and Scott saw us off.

It was a beautiful morning, not too hot, and not too dusty, and by the time we arrived at Bullabulling, about 18 miles, we were well prepared for our breakfast, which we did ample justice to. Here we met a Mr Nauman and Mr William Thompson — President of the Chamber of Mines — returning to Coolgardie. At 2 p.m. we arrived at Woolgangee, the horses dead beat, in fact we had to get down several times and walk. Had we not, I doubt if they would have been able to complete the journey. It is no joke walking by the side of the coach, your feet going deep into the sand at each step.

Woolgangee 44 miles from Coolgardie.

When nearing Woolgangee, we saw the first sign of rain having visited the locality. Water was standing on the track in places, the very sight of it cheered the breasts of everyone. No rain had fallen for months, and only a limited number of teams were allowed to leave Woolgangee for Coolgardie, there being scarcely any water to be obtained on the route. What water they had at Woolgangee was being brought by the railway, in huge tanks fitted on the railway tracks, from Northam 244 miles, but not half enough to supply the wants and requirements of the place.

At Woolgangee we dined, not well, but sufficiently, thousands of flies to keep us company, the walls, ceiling and tables black with them. At 4.30 the long looked for rain came down, a most terrific storm, accompanied by thunder, lightning and hail, burst over the place. Just before the storm commenced, a willy-willy passed over the spot where we were waiting for the train. At its approach

Willy-willy (a dust storm) called in the Nor'west a "Cock-eyed-Bob".

everyone ran to the nearest shelter; we disappeared under the platform. I had never witnessed such vivid lightning; fortunately, as the rain and hail commenced, the train from Southern Cross, which we had been waiting for, arrived and we were enabled to take shelter in a carriage.

In half an hour the whole place was one sheet of water, the dam being filled to a depth of about 5 feet. The dam is formed to conserve the rainwater — at the foot of a granite blow, or low hill, dwarf walls being constructed round it, by these means the water is conducted to the dam.

This dam held 1,367,700 gallons.

Directly the rain ceased the entire population turned out, with every imaginable receptacle for holding water. You never witnessed such a busy sight — men, women and children with tanks, pots, pans, pails, jugs, saucepans and every description of article baling up the precious fluid, which they'd not seen in such quantities for months.

We left them at it when the train moved off for Southern Cross, where we arrived at 10.45. This journey is only 74 miles, but the line not being properly made up, that is, metalled, the progress was slow. It occupied nearly six hours, at least four of which were passed in the dark — there being no lights in the carriages. On arriving at the "Cross", we enquired when the train started for Perth — at 11.45, we were informed, just time enough to rush to the Hotel to get something to eat. We had had nothing since 2.15, so we wanted food and drink badly. We got that, but we missed the train by one minute — the train had gone before the time stated. I'm afraid somewhat strong language was used on that occasion, however nothing was to be done but to return to the Hotel — full up, no beds. I slept on a sofa, Bobby and a man with us on the floor of the sitting room. Before we turned in, our companion, whose life we had nearly saved, stood a bottle of champagne. We had known something of him at Coolgardie, he was in the office of one "Viner", a solicitor, a ruffian of the worst class — he had defended Lowden — besides which he owed our Company a sum of money, advanced to him by Lowden out of money sent to work the mine, but he declined to repay it. Before leaving however, I had the opportunity and satisfaction of telling him my opinion of him, which he strongly objected to, and said, "I'll make you pay for saying that", but as the interview took place at 5 o'clock on Boxing morn, in the "open" with no one near — Well!! He had no witnesses. But to return to our companion — we met him at Woolgangee, he was ill, very ill, and awfully low-spirited. We gave him some fruit we had purchased, and cheered him up. He travelled with us to the "Cross", and before he went to sleep he was quite another man. I should say he blessed the moment when he met us. We also met at Woolgangee Mr Varden, his wife and child. Mr Varden was the manager of the Brown Hill Mine — he was not well, and was on his way to Albany to recuperate. It is a trying climate, and very few can stand it for long.

Mr Briden.

The next morning we were up early, had a good

breakfast and left by the 7.35 train for Perth, where we arrived at 9.45 p.m. — 14 hours and 10 minutes. 248 miles — about 18 miles an hour. From the "Cross" I brought away a glass — presented to me by the Landlady — as a specimen of those used for 6d. drinks. I had brought another from Coolgardie, but there, as I have said before, they are 1/- drinks — they are only curios, on account of their size, they don't hold much!!

At Northam the train stopped sufficiently long to enable us to get an excellent dinner, at really a good hotel, and at Parker's road we purchased fruit and biscuits.

After leaving our luggage at the Grand Hotel, we walked up Barrack Street to the Oyster Shop. We did just enjoy those bivalves, washed down with brown stout, made us think we were in dear old England.

They were quite pleased to see us again at the Hotel, gave us the best room in the house, the window opening on to a wide covered balcony; as it turned out, it was fortunate we had that room.

Claremont — road to the Osborne Hotel.

Friday January 10th. As we were returning from sending a cable, we met Dr Davy, who invited us to dine with him that evening at the Osborne Hotel, of which I have already given a description. After lunching off oysters, and looking round the City, we started for Osborne — the station being Claremont. Somehow I think I went to sleep, and Bobby was thinking and we passed the station, but we got out at the next, this gave us a walk across some rough country, which — bar the flies — we enjoyed.

About Flies.

What are the good of flies? They are an intolerable nuisance, no matter where you drive into the bush, far away from the haunts of man, there they are — if they were only like what I might call the "Domesticated British fly" it would not be quite so bad, but these wretched flies set no store on their lives, they get in your eyes, that is the corner of your eyes, and are only removed when killed.

Dr Davy and his family were inhabiting a bungalow near the Hotel where we dined. His wife and another lady made up the party; it was a good dinner, coffee and cigars on the balcony after, it was a lovely cool evening, after the heat of the day, remember it was summer, although in the middle of January. I was not then feeling well, not up to the mark, but I didn't think I was so near being seriously ill.

The following day (Saturday) Dr Davy lunched with me — Bobby had gone to Fremantle for a sail. We had oysters and Chablis, then strolled round the City, and unfortunately I ate some unripe peaches, they looked so tempting, and it was very hot. In the evening I dined alone, then looked in at "Cremorne", where I smoked a cigarette, and had a whisky and listened to the Band and some songs.

Sunday morning at 1 a.m. I was taken ill, very bad pain. Early in the morning I went out to find a Chemists, no shops open and no chemist to be found. The landlady came to my rescue, and gave me something, which brought relief for a time, and the chambermaid (an old one) acted as nurse, and applied hot flannels and other remedies. In the afternoon Bobby returned from Fremantle. The next morning he went for Dr Davy, who pulled me through. I did not leave the room for 9 days, except to go out on to the balcony which, being partitioned off from the part used by the Hotel visitors, I had to myself.

During the time I was laid up, I had many callers, and every one, down to the Japanese Cook, showed the greatest sympathy, the latter even to the extent of making a large bottle of the strongest beef tea for me to take during the journey to Albany.

Monday January 20th. I walked out of my bedroom into a carriage, the first time I had been out, and caught the 3.30 through train for Albany — 351 miles — where we arrived at 11 o'clock the following morning. Many friends saw us off, including Mr Horgan, Mr Barnier and several others, who came to wish us a quick, and safe return to the old country.

Twenty hours in the train was a long journey when not up to the mark. However, I was so many miles nearer home, and at Albany I was in sight of the sea, which was a joy indeed.

After leaving our things at the Royal George

There was a very nice motherlyest of women at the R.G. Hotel by name of Mrs Dunn. She said she was waiting for her husband John G. Dunn, the discoverer of the Wealth of Nations, Coolgardie, who was coming home from Adelaide. She made a pair of slippers with forget-me-nots worked on. She told me they were for her husband.
She went to meet him on the tender, found him with a fluffy-haired bar-maid. There was a frightful row.
It turned out she was not his wife. She'd been a lodging house keeper with six children – had deserted her husband for this man Dunn.

Hotel, the same we stayed at when we arrived in the country, and leaning on Bobby's arm — I was fairly weak — I walked up the jetty, the very smell of the sea seemed to revive me. I weighed myself on the jetty, turning the scale at 11 st. 12 lbs — my weight when I arrived in the country — by the same machine — was 13 st. 11 lbs. So I had lost in weight 1 st. 3 lbs.

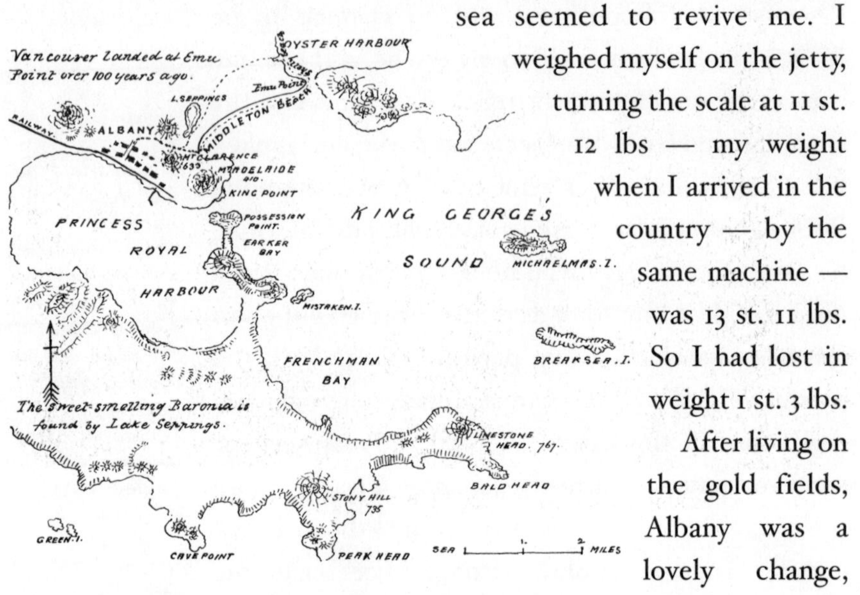

After living on the gold fields, Albany was a lovely change, like being at Dawlish or some other part on the Devonshire coast. It is landlocked, surrounded by hills. In front of the main street, in which our hotel was, is Princess Royal Harbour — about 4 miles long by 2 miles wide (sea miles). On the opposite side of the Harbour are hills covered with verdure; to the left, towards King George Sound, are hills and rocks which stay the inroads of the ocean. The entrance to the harbour is somewhat circuitous, but the channel is sufficiently deep to permit the biggest ships to enter. I believe the whole of the British Fleet could ride at anchor in this harbour.

1906. One of the grimmest ironies of luck is exemplified in the fact that John G. Dunn, the discoverer of the famous "Wealth of Nations" was last week lodged in the Coolgardie lockup to serve a months imprisonment for a paltry debt of £20. (Western Mail, March 31.)

The day after our arrival, we took a most delightful drive to Middleton Beach, about 3 miles

Albany, looking towards the entrance to Princess Royal Harbour.

from the town. It was simply lovely. Imagine a beautiful bay which appeared to be surrounded by high hills covered in most parts with fine trees, and shrubs, the water a pure emerald green, the sand perfectly white, stretching for two miles in a straight line, the road finishing abruptly on the sands. From this point we had a glorious view, and for some minutes we stopped, enjoying the magnificent prospect. It appeared to be a lake, the entrance to the bay from the ocean not being visible, but as we drove along the sands, it gradually came in view. We could see the huge waves in the far distance rolling into the bay, and, as they advanced, losing their power, ultimately degenerating into a gentle ripple. It

Albany, from the east.

was indeed a grand sight. In driving along the sand we found it necessary to keep in the sea, where the sand was hard, above sea level it being so soft our horse objected to it. The length of the beach was quite two miles and appeared to end at a hill, but when we came to Emu Point, we found the sea passed through a channel into another bay. This channel was about 500 yards wide, and some men were bringing across horses and cattle. We watched this with great interest. There were four men with oars, and one at the end of the boat who had a rope with a noose at the end. This they placed round the neck of the animal to be brought over, he was then drawn into the water and had to swim across. We

This bay is called Oyster Harbour.

met some people at this point having a picnic, and they told us of another way back through the bush, and a most delightful drive it was, nothing but a bush track, part of it running by the side of Lake Seppings. We saw some large shrubs with a red flower growing below the leaves, more in fact like a fringe than a flower, they were very curious. We also saw many other fine shrubs and trees, different to any others we had hitherto seen in the Colony.

The sweet smelling "Boronia" is found at Lake Seppings.

The next day was cold and windy. Sir Joseph Renals (late Lord Mayor of London) and his wife arrived in the *Arcadia*; he presented an "Interviewer" with a box of 50 cigars, which were so bad be was unable to smoke them.

The next day was beautiful. We called on a well known character, a collector of seeds, birds, animals, reptiles and insects. He very much wanted me to purchase a fine young eagle, 3 months old. It stood 2'6" high, but I contented myself with purchasing some seeds, a few birds — dead ones — and an opossum. He told us it was rather rare to get an eagle as they were being shot off, owing to their liking for young lambs.

Albany, from the north.

We then spent some time on the

Royal George Hotel, Albany.

jetty watching the boys fishing. They caught a number of skipjacks, a very good fish to eat, and "leather jackets", which are not good to eat — they cut these up for bait, a cruel process.

I also spent a considerable time in looking for sharks, I very much wanted to see one, but no such luck. They have a good number in the Harbour, and precautions are taken by enclosing the bathing place with an open fence. The sharks can only have a "look in", nothing more.

Saturday January 25th. Our last day in "Westralia". The rest, coupled with the sea air of Albany, had done me a great amount of good, and I felt considerably stronger. A long time was occupied in packing; Bobby's trunk had been brought from the custom house, where it had been since our landing in W.A. In it were most of the things we required for the voyage. My trunk had been at the Grand Hotel Perth, so that our things had got what you may call mixed. During our stay at the Royal George Hotel, we had been very comfortable, and I should recommend everyone to put up there, the Landlord and Landlady being most obliging, and the position of the Hotel facing the Harbour being first class. It is also quite near to the Railway Station and the Jetty. We met some friends, who were pleased to see us again — Mr Varden of "Hannans Brown Hill", and his wife; Mr Stoddart, a fine representative of the Scotch race, and a real good fellow, he is now a J.P. of Coolgardie; and a Mr Rendall, a pleasant sociable man, a J.P. in England, but in Australia a miner. He was part owner of a mine at Londonderry, and had his wife and family living in a "humpy" (a shanty or small house) in Albany.

Mr R Emeric Tyler, F.R.I.B.A.

Well! After many parting drinks, we said goodbye to them, and others we had known on the "fields" and at midnight went on board the R.M.S. *Orient*. She had steamed into the Harbour about two hours before. When we saw her lights, we said, "Now for England, home and beauty — and a long farewell to the land of gold, sin, sand, sorrow, sore eyes, flies and shilling drinks".

The R.M.S. Orient.

Postscript

In spite of Bobby's farewell comments as a postscript to his father's letter of 30 December 1895 — "Hurrah! We are leaving this land of sand, sun and sorrow" — after studying mining engineering in London he returned to Western Australia.

He returned as manager of the Murchison Associated Goldmines Ltd mines of Rubicon and East Fingall at Day Dawn, a small mining settlement near Cue.

At Day Dawn he met and married Elizabeth Georgina (Bessie) McFarlan, daughter of George Robert McFarlan, a naval captain, in January 1902.

In 1904 Robert Snr and his dear Emma visited Bobby and Bessie at Day Dawn, and a letter from Emma to daughter Lulu (and her husband Ernest) back in England, although not part of the original diary, is reproduced here for the insight it gives into early Western Australia and Bobby's new life on the Goldfields.

The Rubicon Mine,
Day Dawn.
April 7th 1904

My Darling Daughter,

Today is your birthday. I thought of you the minute I woke, and Pa is sending you off a cable to wish you ever so many Happy Returns of the Day — and that your new life may bring you much joy and happiness, as free as possible while here below from troubles, cares and anxieties. We shall drink yours and Ernest's health tonight most sincerely and heartily. I hope Ernest has given you the sovereign from me for your birthday present and that you will have spent a very happy day.

Well now, dear, to resume our journey. I left off a day before getting to Fremantle — it was rough and I was so sick I could do nothing until the ship got into the harbour at 5 a.m. Then I jumped up and packed my poor clothes as well as I could and I am afraid they will all be spoilt, when in the middle of it all in walked my dear boy, looking so well and so delighted to see us. He had breakfast with us on the ship and we all left with compliments paid and good wishes from all and cards and invitations everywhere. Then we took the train to Perth, getting there about 10 a.m., and on to the new Palace Hotel. We had been up since 4 a.m., but now the heat began — and Oh me it was dreadful! I was just prostrate and could do nothing all the afternoon. Bobbie took possession of Pa and they went off calling and interviewing. We dined at 6.45 p.m., and then went to see the play "Gay Parisienne", and very good. It was a treat for Bobbie.

We had to be up the next morning at 5 a.m. to catch the train to Geraldton. I was feeling rather bad — the heat is so awful. Oh, I have forgotten, we had to spend Good Friday in Perth because there were no trains running so that I passed a quiet day comparatively and in the afternoon we went for a beautiful drive all round Perth, calling on Mr Morgan. All the morning I was arranging our luggage and oh, the heat was awful, directly I move the water runs off me. I must get thinner.

By 8 p.m. I was getting to feel so bad I had to go to bed, especially as I had to leave so early in the morning. By the time I was dressed on Sat. morning (5 a.m.) I had the commencement of an attack of colic and sciatica all down the left hip and leg — could hardly walk. However I managed to get to the train and Bobbie was so good to me. We had a large hamper and a nice carriage with lavatory all to ourselves — so my spirits revived. But we did not arrive at Geraldton till one in the morning! Fancy — from 6 a.m. till 1 a.m. I shall never forget that but we were so happy to be together (we three), talking and telling everything — it made up for a great deal. Then we went to bed and again, being Easter Sunday, we had to remain (no trains) and I was thankful for the rest, but could not get rid of my pains. I had some neat brandy and that made me better and sat on the verandah of the hotel all day, resting and reading, thinking and watching the sea — which is very beautiful here — but, oh! Such a beggarly town and the heat — 90 degrees in the shade and the people saying it was a good cool day!

Well, we went to bed early and the next morning we were up at 4 a.m., this time — train leaving at 5.30 a.m. for Day Dawn. We had another hamper and another carriage to ourselves all the way and arrived at 9.30 that night. But, oh! As the day passed and we were travelling up the country away from the sea, the heat became worse than ever — over 100! I shall never forget it. Pa did not feel it nearly half so much as I did — the wind blew hot — the sun blazed and not a cloud in the sky so perfectly blue. Nothing to be seen but pretty trees on red sand and miner's camps spread here and there. At the stations the miners with just trousers and shirts on, and I think they looked upon me as a curiosity. Tea at every station and Bobbie paying me every kind attention — but oh! I shall never forget it. The miners said it was pretty warm! Now and again miners' wives and children, prettily and in a way, fashionably dressed, and I suppose able to bear the heat, for they looked quite comfortable. Well, on and on we went (I saw a large kangaroo!), and at last we arrived. The evenings are a little cool and so I revived. Two buggys and a cart to meet us and several gentlemen, among them being Mr Wills (very nice), Mr Stopher, Mr Ferguson — all seemed delighted to see Bobbie back. Mr Wills drove me in his buggy out into the darkness to the mine — most polite, apologising for the rough driving. Pa, Bobbie and Mr Ferguson in the other buggy and all our luggage in the cart. Bessie met us at the camp and Bobbie brought us in with great delight. Supper was waiting for us, all very nice and we were then shown into our bedroom — which is a very nice one indeed with every comfort. But the journey was very hard and the heat on the train so bad — on the mine the heat was 100 in the shade. Of course I got ill and Tuesday and Wednesday I have been in bed with a bad attack of colic — a kind of dysentery. I felt very sad and frightened and I could only think of all at home, but Bobbie gave me chlorodyne and the rest and the cool that comes in the evening has revived me and I am getting quite well again and feeling myself. If only the heat goes down I hope to get quite well before I start again on our trip.

Bessie has been very kind, can't do too much for me and I am living on iced soda and arrowroot. So don't worry about me at all. Mr Stopher, who has just come here from Vancouver, says that the worst of the journey is over, for he felt the journey here from Perth and the heat here worse than any other. He is a kind young man and Bobbie likes him but he is no help to him at all for he can't manage the accounts and gets into many muddles — but he knows it and is most anxious to please and thinks himself lucky to be near Bobbie and Pa. He comes up to the bath here every morning. Pa is wonderful — he has been down the East Fingall mine, twice down the Rubicon and in parts where Mr James himself has never been — drives round every afternoon with Bobbie interviewing people and is most highly thought of by all. Bobbie is so proud of him. When I get quite well I am to be driven to see all the camps and many callers are coming on Sunday to see me.

The little three year old, Gracie, is a dear funny little thing, you can't help liking her and of course, Pa has taken a great fancy to her and vice-versa. He pets and spoils her. Bobbie is most funny with her — she is so frightened of him and so obedient. Bessie does not spoil her at all. The other girl, Doris, is a good looking, refined, ladylike girl — 14 yesterday and is quite the waiting maid at everyone's beck and call and Bessie is not too gentle with her. Altogether I feel sorry for her (very); she has the charge of her little sister too. Bessie had a servant, but a very bad one, not a good cook and Bessie has to superintend everything and of course gets worried for Bobbie is not too patient and he wants his father and me to have everything of the best. The soda water, Chablis, ale, etc. that is consumed here would astonish you, but in such awful heat it is utterly impossible to do without it. I drink soda or milk or lemonade — Pa, soda and Chablis (iced). I can see Bobbie keeps a most hospitable camp and is very popular with them all.

Your photos of the wedding group are most admired — Bobbie has chosen the sepia one and Mr Horgan is to have the other. The bridesmaids are much admired, especially Ida and Bessie. Bobbie and Pa are delighted with the cake and the presents — also the cigarettes Ernest sent and thanks you very much for them. Bobbie works very hard and it is just marvellous to me to hear him ordering and deciding, talking mining matters with the men in the most masterly way. I can't think how he has managed to learn it all.

Tell Nellie that they are very pleased with the electric light machine tho' it does not work very well yet — and tell Mrs Frank that Bessie sends her love and thanks her very much for the books and texts. The texts are now in their bedroom and the testaments are all received most gratefully.

Next week Pa and Bobbie are going round Coolgardie and Kalgoorlie and I remain here — then we will all go to Perth for a time and then we part. Bobbie cannot leave the mine to go with us to the eastern colonies and cannot afford the money. Now I am better I am very glad we have come here — it is such a pleasure to your dear brother. He often says how he should have liked Ernest and you to have been with us. I wonder whether you went away for Easter? I fancy you said something about going to Felixstowe. I do hope dear, you are both well and that all is going well at home. I have not yet had your letters here, but I suppose I shall get them on Saturday in Perth. Please give my love to everyone. I can't name them all. I know when I am home again I will amuse them with my various experiences. Bobbie says I am looked upon as a "curio" out here and considers Pa quite a "dude". Your should have seen Pa when he came up from the mine — you wouldn't have known him!

I think now I have told you everything up to date. The country round here, the space is grand, the sky always without a cloud, the sunsets magnificent and always a strong wind mornings and evenings which make it so much cooler. We live in the air — the wire netting doors keeping out all the flies and animals — 3 cats, a kitten, dog and puppy and 17 chickens. We get lovely eggs.

Many kisses dear to you and fond remembrance to Ernest and love to all — from your Everloving mother,

Emma Tyler.

Bobby Tyler, c. 1900.